Ragstone to Riches

Imperial Estates, *metalla* and the Roman military in the south east of Britain during the occupation

Simon Elliott

BAR British Series 638

2018

Published in 2018 by
BAR Publishing, Oxford

BAR British Series 638

Ragstone to Riches

ISBN 978 1 4073 1652 9

COVER IMAGE *Artist's impression of Roman villa at Teston in the upper Medway Valley. Watercolour by Mac McKeown.*

Printed in England

BAR
PUBLISHING

BAR titles are available from:

BAR Publishing
122 Banbury Rd, Oxford, OX2 7BP, UK
EMAIL info@barpublishing.com
PHONE +44 (0)1865 310431
FAX +44 (0)1865 316916
www.barpublishing.com

Acknowledgements

Many individuals have contributed to the completion of both *Ragstone to Riches* and the preceding PhD thesis, and to these I would like to give thanks. First and foremost, Dr Steve Willis at the University of Kent for his belief in the project. Secondly, to the academic team at UCL's Institute of Archaeology who facilitated the initial Roman occupation-period Medway Valley research, particularly Dr Andrew Gardner. Also Professor Andrew Lambert at KCL's War Studies Department for his ongoing counsel and mentoring. Finally, the host of leading experts in their respective fields who have kindly given me their time and thoughts in personal interviews to support the research – Dr Paul Wilkinson on the occupation-period north Kent coast and Medway Valley, English Heritage's Tony Wilmott on hydraulic river infrastructure, Dr Gustav Milne on the *Classis Britannica* and Roman maritime technology, Dr Malcolm Lyne on Roman pottery in the South East during the occupation, Jeremy Hodgkinson on the Weald during the occupation, David Rudling on East and West Sussex during the occupation, the British Museum's Sam Moorhead on numismatic insight into Kent during the occupation, the Museum of London Archaeology's Ian Betts on occupation period tile manufacturing, the Museum of London's Meriel Jeater on the Roman walls of London, Professor Sir Barry Cunliffe and Professor Martin Millett on their respective appreciations of Kent during the occupation, my cartography consultant Elizabeth Blanning for transforming my hand drawn maps into the beautiful plates presented here, and finally my proof readers Dr John Lambshead and patient wife Sara. All have contributed greatly to enable this research to be as up to date in its approach and conclusions as is possible. Thank you all.

Contents

List of Figures

List of Tables

List of Maps

Preface

This work considers the industrial scale extractive industries (*metalla*) of the south east of Britain during the Roman occupation. These were the iron industry in the Weald which catered for the demands of the whole province (including the large-scale military presence in the north), and the ragstone quarrying industry in the upper Medway Valley which provided vast quantities of ragstone to build much of Roman London. While much is known about the former, little was known about the latter until this research began. 'Ragstone to Riches' therefore presents a new synthesis regarding the iron industry, and then – most importantly - brand new data and analysis concerning the ragstone quarrying industry. The latter includes the identification for the first time of the five specific quarries utilised during the occupation, and the economic model used to transport the extracted stone to its place of use. It also shows for the first time how the two industries were intrinsically linked until their mutual demise in the mid-3rd century AD. The work also discusses whether they were industrial Imperial Estates (singularly or jointly) owned by the Emperor, and considers the role of the Roman military in their management and operation.

The research is based on my PhD thesis at the University of Kent, completed in 2017, which was titled *'Change and continuity in the exploitation of natural resources (such as stone, iron, clay and wood) in the principal areas of industrial activity in Kent (namely the Weald, Folkestone region and upper Medway Valley) during the Roman occupation'.* This concluded that the greensand quern-stone manufacturing industry at East Weir Bay in Folkestone - part of the original research - was not a *metalla* (the Roman term for their extractive industries) of the same scale as the ragstone quarrying and iron industries detailed above, and so it is not considered here.

List of Abbreviations

ASE	Archaeology South-East
BDHS	Battle and District Historical Society
CLBR	Classis Britannica
DAG	Dover Archaeological Group
FRAG	Folkestone Research and Archaeological Group
HAARG	Hastings Area Archaeological Research Group
HER	Historic Environment Record
IHRG	Independent Historical Research Group
KAFS	Kent Archaeological Field School
KARU	Kent Archaeological Rescue Unit
KAS	Kent Archaeological Society
LIA	Late Iron Age
LMARG	Lower Medway Archaeological Research Group
MAAG	Maidstone Area Archaeological Group
MOLA	Museum of London Archaeology
MOLAS	Museum of London Archaeology Service
SAS	Sussex Archaeological Society
SEAS	South East Archaeology Services
VCH	Victoria County History
WIRG	Wealden Iron Research Group

Finely worked upper Medway Valley Kentish ragstone blocks in the land wall of Roman London (this location near Tower Hill), late 1st century AD. Hall and Merrifield (1986, 28) said over one million squared and dressed ragstone blocks, from the five quarries identified for the first time in this research, would have been used for the inner and outer facing alone. Simon Elliott.

1

Introduction

Developments in the exploitation of natural resources such as iron and stone by human society can be a key indicator of change and continuity in societal priorities and practice. In so doing they provide a forensic level of insight into the changing daily lives of communities who lived in the past. A particular case in point is Britain during the Roman occupation when, for the first time in the long narrative of human occupation of the islands, such exploitation began to occur on what we might recognise today as an industrial scale. The degree of human and environmental impact this created far exceeded anything that had gone before, or indeed afterwards until the advent of the Industrial Revolution in the 18th and 19th Centuries AD (Laurence, 2013, 43, and S. Elliott, 2014b, 49). Crucially, in the South East (defined here as Kent, East Sussex and Surrey given their importance to the occupation period regional extractive industries) we can detect this development through many diverse datasets (many newly available through this research), and also appreciate it through the use of anecdote where academically appropriate. To that end, the core aim of this book is to use this data to understand and interpret iron manufacturing in the Weald and ragstone quarrying in the upper Medway Valley during the Roman occupation, and then to consider the commonality or uniqueness of this experience when compared to elsewhere in occupied Britain and indeed the wider Empire. This includes a detailed evaluation of whether these were industrial Imperial Estates owned by the Emperor, and the level to which the military managed and operated them.

This book is based on the key data set out in my PhD thesis at the University of Kent (Elliott, 2017), and my earlier MA Dissertation at UCL's Institute of Archaeology (Elliott, 2011). The latter, which set me on the path to the PhD research, revealed the intense level of industrialisation that occurred in the region during the occupation and how this changed over time. Hitherto, this evidence had not been collated and examined synthetically. A particular focus of the MA research was ragstone quarrying in the upper Medway Valley, both at the tidal reach around Allington on the River Medway and also, in an even more intensive way, above the tidal reach. The MA hypothesis also discussed whether hydraulic riverine infrastructure would have been needed to make the river usefully navigable above the tidal reach to enable the ragstone quarries there to be exploited to their full economic potential through easier transportation (see Figure 1 for evidence of this maritime transport network, and Figure 2 for detail on the likely Roman transport vessel types used). I also discussed the potential for an association between the various known villa estates on the river and the elites who ran the quarries, using available datasets to show these connections where they could be shown to exist. Dubbing the work the 'Medway Formula', I concluded by arguing that a reasonable case could be made that the natural resource exploitation through to the middle of the 3rd century AD in the upper Medway Valley[1] was facilitated by the *Classis*

Figure 1: The 'Medway Stones' – millstone or column base blank, one of four which the author helped recover from likely Roman wreck site in the River Medway, possibly Greensand from East Cliff, Folkestone. Simon Elliott.

Figure 2: Mosaic image of Roman merchant vessel, Ostia Antica, Italy. Transport vessels of all sizes played a vital role supporting the *metalla* of the south east of Britain during the Roman occupation. Simon Elliott.

[1] The River Medway and its valley feature heavily in Chapter 4 and were earlier the specific focus of my MA research as detailed above. A number of definitions are available for the various sections of its length. For example the Environment Agency (EA) divides the river into four sections for its flood warning system. However, for the purposes of this research I believe that a simpler system will be more useful. To that end I will use the definition utilised by the Victoria County History's 'England's Past for Everyone' Lower Medway Project (carried out in association with English Heritage, Hann, 2008). This defines anything below the tidal reach at Allington as the lower Medway, and anything above it as the upper Medway.

Britannica regional navy on behalf of the state (at least in terms of transport), after which it became much more localised in management if, nonetheless, still being largely geared to the demands of Imperial projects.

The PhD research built on the MA work with a view to determining if the 'Medway Formula' findings of intense resource exploitation could be replicated across Kent and the South East, and more broadly across occupied Britain. The interpretative path along which this research ran has now provided the template for this book. To make it as accessible as possible it is broken down into six chapters. This first is this Introduction which also features background on the specific research questions to be asked and the key research parameters. The second chapter is a general background section designed to be of common use to all of the following specific areas of the research, and to allow a view to be taken regarding systems of authority, exploitation, power and agency in occupied Kent and the south east of occupied Britain. In so doing, it covers the current debate on the nature of Roman imperialism and its likely manifestations in the region in terms of social structure and systems of control. Specifically, this background material will feature:

- A review of the geology of the South East and its impact on the economy and settlement in the region during the Roman occupation, important given the focus of this research on change and continuity in the exploitation of natural resources.
- An overview of the latest thinking regarding the workings of the Roman economy to provide context for the later regional analyses (this including a general background sub-section which considers the extent of market integration across the Empire, a forensic examination of the nature of the Imperial economy, a similar examination of the nature of the provincial and market economy, and finally a detailed analysis of Imperial Estates given their importance to many aspects this research).
- A broad review of the Kentish experience of the Roman occupation, given the county's importance to the research.
- An analysis of Roman industrial practices to provide a backdrop for the focus on the occupation-period extractive industries that were exploiting the natural resources.
- A discussion on occupation-period maritime transport in the South East to again provide context for the regional analyses.
- An examination of the military presence in Britain during the occupation with an eye to their relevance in the wider research.

The third and fourth chapters are the central part of the research and comprise two specific regional analytic studies, their order deliberately based on the maturity of their respectively available datasets, with the best-known first. To that end I begin with the Weald and its comparatively well-documented occupation-period iron and tile/ brick industries. These have been examined and interpreted over a lengthy period of time (for example by Ernest Straker, Gerald Brodribb and Henry Cleere, the pioneers of Wealden archaeology in the 20th century) though in practice the evidence and synthesis has been patchy and there is now much new evidence which is presented here. In Chapter 4 I then revisit the 'Medway Formula' where newly available data from my PhD research has allowed a much more detailed picture of the occupation-period Kentish ragstone quarrying industry to emerge, together with associated settlement (for example the villa at Teston illustrated by Figure 3 and the regional Roman road network referenced in Figure 4). Both regional studies begin with a review of the details of all of the relevant key primary evidence sites, providing analysis set against the dating of the location, nature of occupation and the type of site as presented by the available evidence, all threaded through with the available bibliographic detail recording the history of the site. Given the focus on the exploitation of natural resources and the associated extractive industries in this book, specific comment is also made on the scale of the industrial sites in both regions, both in the site summaries and main body of the research. Where appropriate analogous modelling is used here, for example Peacock's hierarchy of modes of production regarding the Roman pottery industry (1982, 8), where the linkage between capital and labour at the larger sites he modelled is particularly useful when reviewing Kent and the South East's extractive industries during the occupation (see Section 2.4).

Figure 3: Villa excavation at Teston, 2014 season. South wall of main range at top, later buttress centre, apsidal with plunge pool foreground. Simon Elliott (Site Co-Director).

Figure 4: Roman road running from Dean Street Roman quarry to Roman ford at Barming/ East Farleigh villa. Two burial groups and milestone/ burial marker along its length. Located and examined by author. Simon Elliott.

Next, the fifth chapter features a detailed discussion reviewing the data and evidence presented in the two regionally focused chapters. It is here that the provenance of the two considered *metalla* as Imperial Estates, and the role of the military therein, is considered. The sixth chapter is a short conclusion which synthesises all the preceeding data and interpretations to present an overall picture of the *metalla* of Kent and the south east of Britain during the Roman occupation.

Three appendices follow which provide a reference resource to the preceding chapters. These include a summary of all of the regional natural resources exploited for use in building and other activities in occupation-period Kent and the South East, a discussion on the sources of data used when referencing the *Classis Britannica* which is so important to this work, and finally an analysis of Roman riverine maritime technology given its importance to the transportation of extracted natural resources in the county and South East during the occupation. The book closes with a list of abbreviations and a bibliography.

1.1 Research Questions

For my PhD research a number of related subsidiary themes and hypotheses were tested given their value to the wider research, and these are rehearsed in this book given their value to the wider work. These include:

- An examination of the hypothesis that towards the end of the occupation settlement in Kent and the south east of Britain deteriorated significantly, leaving two 'islands of activity' in north west Kent (centred on the Darent and Medway Valleys, still economically orientated towards London even at this late stage) and eastern Kent (centred on Canterbury, Roman *Durovernom Cantiacorum*, and the Channel coast). This utilises research new to this work, and also an examination of existing published research.
- Whether the experience of Rome in the province[2] would have been at its most undiluted in Kent and the South East, especially early in the occupation, given the region's proximity to the Continent. Blanning's (2014, 484) assertion that Roman Kent had more in common with northern Gaul than elsewhere in occupied Britain will be considered in this regard.
- The conception that agriculture increased even further in importance in relation to the local economy later in the occupation in the region, in the context of a discussion regarding the decline of state-run extractive industrial enterprises (if that is so determined) and increased demand from the Continent for food and provisions.

Finally in this sub-section, note that any discussion regarding theoretical approaches to the study of the Roman Empire and its economy (for example the debate regarding levels of Imperial exploitation in occupied Britain, and the concept of Romanization and its more recent counter-arguments), can be found in Chapter 2, and particularly Section 2.2.1.

1.2 Research Parameters

Next I make the following general points regarding the research parameters used in the book.

- Within the reviews of primary data in the two regional surveys, all sites (excepting a few examples such as isolated temples, cemeteries and bath houses) have been identified as one of four specific site types, namely villa settlements, non-villa settlements/ 'native settlements', small towns (these three based on Richard Hingley's tripartite system, 1989, 20), and industrial sites (the latter added by me given the focus of the wider research on the extractive industries). As set out by Hingley (1989, 21) villa sites are determined in the research as domestic buildings showing evidence of the investment of considerable levels of surplus wealth

[2] Where the term province is mentioned, until the Severan/ Caracallan reforms of Britain in AD 211/ 212 (originally planned in AD 197) this references the original single province of Britannia. After the reforms it references either Britannia Superior or Britannia Inferior dependent on the context (for example in the case of the Classis Britannica the former, see Section 2.6.2, and in the case of the northern frontier the latter), or indeed where appropriate both. Following the Diocletianic reformation and the establishment of the diocese with its four and perhaps later five provinces (see Section 2.3.1 for discussion), given the complexity the full name of a given province is detailed for clarity.

in their construction, with Branigan (1982, 81) earlier saying that those building such a structure were very ostentatiously adopting a Roman lifestyle for personal betterment. Meanwhile, I define non-villa settlements/ 'native settlements' (the most common type in Britain during the occupation, Lawson and Killingray, 2004, 20) as those where the inhabitants for whatever reason chose not to display the ostensible trappings of a Roman lifestyle. Meanwhile small towns (defined as distinguishable from their larger *colonia*, *municipia* and *civitas* capital counterparts by the likes of Todd, 1970, 114) are defined in the research as a variety of more extensive, diverse settlements which existed throughout the occupation in Britain and which usually had an association with a specific activity, for example administration, industry or religion (with many also being located at key transport nodes). Finally, for industry I define such sites as those showing activity in this regard but lacking any signs of rural settlement (see discussion regarding industrial Imperial Estates in Section 2.2.4).

- In terms of the interpretative path used when considering hypotheses in the book I have followed the traditional pattern of setting out the data first (this deriving from archaeology, archival research, scientific observation and analogy), then viewing such data through the prism of appropriate modelling, before finally setting out my own views based on the evidence and subsequent analysis (and considering here anecdotal insight where deemed well-judged or prescient). I have chosen to use the widest possible variety of relevant data, ranging from that derived from the sharper end of current peer-reviewed academic research at one extreme, through data available in grey literature awaiting publication or never to be formally published (and made available through initiatives such as the National Roman Grey Literature Project which is being facilitated by English Heritage, Cotswold Archaeology and the University of Reading), all the way through to the antiquarian record at the other extreme. Where there is a potential problem with the data detailed in the sources (for example the interpretation of the military materials found in association with the water mill site at Ickham, Mould, 2010, 144, see Section 2.6.4), this has been specifically referenced and taken into account in my narrative, and as always the use of any antiquarian reference comes with a strong health warning when not associated with modern research (the miss-recording of the 1872 excavations of the Roman villa at Teston being a good example, Elliott, 2013, 40, see Section 5.1.4). In terms of material culture, where the Kent (and other) Historic Environment Record (HER) has been referenced I have provided my own commentary about the provenance of the data. The research also includes the most up to date data and interpretations from my own practical research projects.
- Finally, I take a moment here to present in advance the highlights of the research set out in the book to familiarise the reader such that they are readily recognisable as they appear in the narrative. As will become clear, I have deliberately set out in my work not simply to reinterpret existing data but to provide my own new data through an active field work programme (involving the whole community, ranging from private citizens, local schools, local businesses and local archaeological and history societies, all the way up to regional Government and indeed Government Agencies), thus adding to the wider body of knowledge of Roman archaeology in Kent and the South East, in Britain and indeed across the whole Empire. To that end, I have helped relocate the Roman villa at Teston last recorded (or rather misrecorded) in 1872, and then helped facilitate four seasons to date of excavation. Further, I have found a potential Roman shipwreck complete with load in the River Medway together with its associated quay (a first dive on the site has already been completed by the Royal Engineers on my behalf, with professional dives now planned). Additionally, and crucially for the wider research, I make the case in Section 5.3 below that I have found the quarries from where the ragstone which built much of Roman London and the South East was extracted (fully detailed in Section 4.3.3), and found a new Roman road which linked these quarries with the villa estates where the elites who ran them lived. In association with the new Roman road I have also found a newly located Roman cemetery together with milestone, and most recently an occupation-period iron working site (the first for the Medway Valley). Lastly, a new Roman bridge across the Medway at Tovil may also have been found, again discussed in full below (in Section 4.1.4). As the reader may gather, the new data presented above is heavily focused on the Medway Valley, which has been a particular focus area of my own fieldwork activity. This is reflected in the flow of the book, which deliberately builds from an initial emphasis on background material to inform the later research, then growing to review and re-evaluate existing datasets for the Weald, before moving to the new research for the Medway Valley. This is finally followed by the discussion where the key topics for debate are discussed, thus allowing me to set out my final views and interpretations holistically in the conclusion.

2

Background – Kent during the Roman Occupation

This chapter aims to cover the current debate on the nature of Roman Imperialism and its likely manifestations in the region of study. In so doing it provides all of the necessary background for the regional research in Chapters 3 and 4, and the following discussion in Chapter 5. To achieve this I have broken the Chapter into six topic-based essays which provide a broad view of the following:

- The topography of the region given its importance to the occupation-period economy and settlement.
- The Roman economy, including a discussion on the current debate as to its nature (which considers the levels of market integration across the Empire), then a specific focus on both the Imperial and provincial economies, and finally a review of Imperial Estates (important here given the focus of the research on the regional *metalla*).
- A background overview of Roman Kent and the south east of occupied Britain, with a focus on settlement and the economic historical narrative.
- A review of industrial activity in Britain during the Roman occupation, again to provide context for the research focus on the occupation-period regional *metalla.*
- Given its importance to the regional extractive industries, an analysis of maritime transport in the south east of Britain during the Roman occupation.
- Finally, and again given its potential importance with regard to the regional *metalla* during the occupation, a review of the military presence in Britain during the occupation.

2.1 The Geology of Kent and the South East

Given the focus of this book on the exploitation of natural resources by the extractive industries in Kent and the South East during the occupation, below I provide an overview of Kentish geology and its impact on the economy and settlement in the region during the Roman period. This is further detailed at Appendix A.

The geology of Kent and the South East is surprisingly simple, especially given the wide variety of landscapes evident across the county. It also provides the key to understanding resource exploitation in the two areas scrutinized in this research - the Weald and the Medway Valley. McRae and Burnham (1973, 9) described the underpinning macro-geology of Kent as sedimentary rock laid down during the Cretaceous and Tertiary Periods which has since been forced into a dome, known as the Wealden anticlinorium, by Earth-movements culminating in the Miocene period. This geological process created the two complimentary compound synclines now known as the London Basin and the Hampshire Basin. Subsequent erosion, the development of the Thames Estuary and associated drainage systems (including principal Kentish rivers the Stour, Medway and Darent, for millennia the principal means of accessing the county inland from the north coast), and the formation of the English Channel have then resulted in the pattern of rocks shown on the geological map of Kent today.

These macro geological processes translate into nine geological formations (Croft, Munby and Ridley, 2001, 1-4), which from uppermost to lowermost (generally) are as follows:

- London Clays in the north, which in terms of relief are generally flat and which include the foothills of the North Downs. Lawson and Killingray (2004, 2) explain that much of the surface geology of northern Kent from the Hoo Peninsula to the Isle of Sheppey is formed from these London Clays.
- The Thanet Beds, also flat and comprising of calcareous sandstone, pale grey sandstone and clayey sands.
- Chalk, forming the west-east horseshoe of the North and South Downs, which defines the region. The North Downs comprises the chalk base overlain by extensive deposits of Clay-with-Flint.
- The Upper Greensand, comprising sandy lithoacies originally deposited in stronger currents than the subsequent Gault Clay, hence the separation. This layer is not present beneath the entire London Basin, passing laterally into Gault clay east of a line between Tatsfield and Dunstable, with the Upper Greensand being completely absent on the eastern Kentish coast and thus not featuring as a building material there.
- Gault Clay, on which sits the Holmesdale on the south slope of the North Downs, the most fertile land in the county (Everitt, 1986, 49).
- The Lower Greensand, otherwise known as the Greensand Ridge. This comprises, in descending order, the Folkestone Beds, Sandgate Beds and Hythe Beds which, together with the underlying Atherfield Clay described below, form the geographical region known as the Chart Hills in Kent. The Lower Greensand Formation was particularly important in terms of regional extractive industry during the occupation, for example the coastal outcropping of the very hard Folkestone Beds then being exploited for quern production, while within the Hythe Beds sits the Kentish ragstone which is so important to the Medway Valley research in Chapter 4 (see Figure 5).

Figure 5: Worked ragstone building block from the Roman villa at Teston, extracted from the nearby quarry associated with the site. Sitting within the Hythe Beds of the Lower Greensand Formation, this was the ubiquitous Roman building stone across the south east of Britain until the mid-3rd century AD. Simon Elliott.

- Atherfield and Wealden Clay (sometimes called the Wealden Beds), on which sits the Low Weald and which is the least fertile part of Kent and the South East. The former clay (often described as a sub-division of the Lower Greensand) overlies the latter, they being difficult to differentiate except for the presence of marine fossils. These clays were much used for pottery, tile and brick manufacture during the occupation. Worth noting here is the fact that siderite ironstone mined as iron ore for the occupation-period iron manufacturing industry sits within at least two horizons of the Wealden Clay.
- The Hastings Beds on which sit the High Weald. This geological unit includes inter-bedded clays, silts, siltstones, sands and sandstones. These specific strata make up the component geological formations known as the Ashdown Formation, the Wadhurst Clay Formation and the Tunbridge Wells Sand Formation. These deposits were particularly important during the occupation as within them sat further siderite deposits exploited by the Wealden iron industry (see Figure 6), with the greatest concentration in the whole region being found within the lower levels of the Wadhurst Clay Formation. Additionally, the Fairlight Clay which seems to have been the material of choice used to manufacture the *Classis Britannica* bricks and tiles in the Weald (Peacock, 1977, 237) sits within the Ashdown Formation.
- On the southeastern periphery, the alluvial, wind-blown sand deposits and pebble banks of Romney Marsh. Only a few localities were economically utilized here during the occupation, though since that time large tracts have been brought into productivity through reclamation and aggregate extraction.

This geological sequence had a formative impact on the geography and soils (and thus settlement and land-use) in the south east of Britain during the occupation. In that regard, Everitt (1986, 43) explains that while the changes in geology are less visible when travelling along the modern east-west routes (for example the M2 and M20), they become much more overt when traversing traditional north south drove ways when at least six different landscapes become visible in the space of no more than 65km.

Figure 6: Tap slag from the Roman iron-working site in the Medway Valley detailed in Chapter 4. The siderite which provided the iron sat, in this case, within the Wealden Clay Formation. Simon Elliott.

The above geological formations were exploited to maximum effect in Kent and the South East during the occupation, providing siderite for the iron industry and, principally, a wide variety of building materials. Construction using the latter was one of the key manifestations of *Romanitas* in Britain, as it was elsewhere in the Empire. Examples of the use of stone as a building material do appear in the region in the Late Iron Age (LIA), for example the Ightam Stone used as revetting material and strengthening ballast at the Oldbury *oppida* in western Kent (Thompson, 1986, 270), but it is with the advent of Rome that we see a dramatic proliferation of its use on a grand scale. As outlined by Blagg (1990, 33), demand would have initially been driven by the need for major urban centres to provide the administration required by the new province, and clearly grand stone built structures in this new built environment were intended to put the stamp of Rome on the local populace (especially the elites) as part of a package which included other manifestations of the Roman experience. Tacitus (*Agricola*, 21) is explicit here about the use of stone built structures as part of this cultural suite, saying of Agricola after two years of campaigning in Wales and the north of Britain from AD 77:

> *"Agricola had to deal with people living in isolation and ignorance, and therefore prone to fight: and his object was to accustom them to a life of peace and quiet by the provision of amenities. He therefore gave private encouragement and official assistance to the*

building of temples, public squares and good houses. He praised the energetic and scolded the slack. And competition for honour proved as effective as compulsion. Furthermore, he educated the sons of the chiefs in the liberal arts...the result was that instead of loathing the Latin language they became eager to speak it effectively."

A major early driver here behind this stone-built urbanisation was the societal obligation placed on elites by the state to invest in public buildings in their communities as an expression of their *Romanitas*. Halsall (2013, 90) explains:

"Spending private money on such projects brought important political rewards. It might bring success in the competition to control the curia (town council) of the civitas. The curia was responsible for tax collection... an important source of patronage. Success here... (would)...be a platform for advancement on a broader political stage, the acquisition of Roman, and perhaps even promotion to the higher orders of Roman society."

Such stone built structures, later to be joined by fortifications as the nature of the Imperial experience changed, were expressed in a variety of forms and locations across the south east of Britain, including:

- The provincial and later *diocesan* capital London.
- The *civitas* capital at Canterbury.
- The *colonia* at Colchester.
- The small towns of Rochester and Springhead.
- Conquest and post conquest period fortifications, including the various *Classis Britannica* and later Saxon Shore forts.
- The later fortification of key urban sites.
- The *phari* at Dover.
- Grand stone-built rural elite country houses in the form of villas and similar, from the later 1st century through to the end of the occupation.
- Industrial sites, for example the watermills at Ickham.
- The extensive road network across the region.

Building on this grand scale generated a huge demand for building stone and other materials, creating an explosion in the scale of quarrying together with associated industries such as construction and specialist heavy goods transport. These are all considered in detail in the regional reviews in Chapters 3 and 4.

2.2 The Roman Economy

This work is is primarily a study of one aspect of the Roman economy, namely the extractive industries. Consideration is therefore given here into the nature of this economy to provide context for all that follows. The section is divided into four sub-sections. The first provides background to the theories and modelling of the Roman economy that have been considered over time, including those most up to date. Next I consider the Imperial economy and the means utilised by the Roman state to ensure a province paid its way, crucial to the smooth running of the Empire. I then move on to review the provincial economy, based on markets, free trade and patterns of consumption (Mattingly, 2006, 496), which I broadly refer to as the market economy (in no way implying this functioned in the manner of a modern manifestation of a market economy). Finally I look specifically at one aspect of the Imperial economy, namely Imperial Estates, particularly important to this research given the focus on the regional *metalla* which have often been described as such (for example the iron industry in the Weald, Cleere and Crossley, 1995, 68, and the ragstone quarrying industry in the upper Medway Valley, Marsden, 1994, 83). The description of Imperial Estates set out here specifically acts as a control template for such geographic economic entities later in the research, particularly in the regional analyses at Chapters 3 and 4, and in the discussion in Chapter 5.

2.2.1 Background

The world of Imperial Rome has often been cited as the archetypal 'Empire', attracting interpretations of its strengths and weaknesses by archaeologists and ancient historians from all political spectrums. Central to this debate is the nature of the Roman economy itself, this a vital consideration as it not only helps to explain the fiscal aspects of maintaining an Empire that enjoyed such longevity (though often experiencing times of financial crisis), but also gives great insight into the lives of those who experienced it. A range of views are evident here, from sometimes benevolent to often negative. In terms of the the former, Temin (2012, 2) controversially argues that the quality of life for the average citizen during the Principate was better than at any time before the Industrial Revolution. At the other extreme Pollard (2000, 249), in his analysis of Roman Syria, speaks of an economy where the exploitation of the population through taxation, the immediate requirement to support the army (admittedly this was a border territory, a good analogy for Britain), and the removal of natural resources under Imperial control outweighed the benefits for the population, for example in terms of investment. Similarly, Faulkner takes this negative experience to the extreme in his contention that the Empire, economically and otherwise, was robbery with violence (2000, 120). All of these themes are considered below as I follow the narrative of thinking over the nature of the Roman economy, with a particular focus on how similar to the western pre-modern economy it was (the 'modernists' seeing more similarities, the 'primitivists' less), and how integrated the Roman market was in the broadest sense.

The nature of the Roman economy has long been considered by economists and historians of all leanings, for example Adam Smith, Adam Ferguson, Max Weber and John Kautsky (Bang, 2008, 61). From the 1950s however a model developed by A.H.M. Jones (1953, 293) began to gain ground. This stressed the centrality of agriculture to the Roman economy, he arguing that most of the agrarian produce was consumed locally (with notable exceptions, for example the Egyptian grain supply to Rome, see references to Paul Erdkamp's work below) and emphasising the importance of taxes and rents over trade and industry in the success or otherwise of settlements large and small.

A seminal contribution to this debate was then made by Moses Finley in his 1973 work '*The Ancient Economy*' which, taking the 'primitivist' position, argued that status was a key factor in the economies of the ancient world (Finley, 1999, 45 [original date of publication 1973]). Developing the work of Weber, he argued that these economic systems (including that of Rome) were embedded in social standing, saying that the ancient world placed so much importance in this regard that there was a clear differentiation between the economic activities in which those in the upper reaches of society could participate and those below them. In this way these economic systems of the distant past, he argued, were very different from those of today (at least in the West) where there is a general freedom (he contends) for all to participate in commercial and legal enterprises. He also highlighted the lack of the concept of 'economy' in the ancient world, placing its origins in the early modern world, and also of sophisticated accounting systems (Finley, 1999, 21 [original date of publication 1973]).

An additional and further significant contribution was then made by Keith Hopkins (1985, xiv-xv) who returned to Jones' theme, adding his view that the Roman economic model also allowed for economic growth and was therefore not static. He saw this growth as particularly noticeable in the late 1st millennium BC and the first two centuries AD, with political change in terms of the expanding Empire of the Principate driving surplus production, and the engagement of native populations with *Romanitas* being either indigenously-led or stimulated by the passive encouragement of the Empire (Laurence, 2012, 63).

Next for consideration is Millett's dramatic intervention in 1990 with '*The Romanization of Britain*', a watershed in appreciating not only the Roman economy but also Roman Britain, with its principal focus on cultural change. His work had its earliest roots in Francis Haverfield's 1906 '*The Romanization of Roman Britain*', it then being incubated through Finley's conception of the Roman economy, and the subsequent academic debates of the 1970s and 1980s regarding the nature of this economy. Using socio-economic and anthropological models to analyse a wide range of archaeological, epigraphical and historical data (though with an emphasis on the analysis of material evidence), his central contention of the Romanization process was that the integration of a province into the Roman Empire (in this case Britain) was a two-way social acculturation process, with complex patterns of interaction between the incoming world of Rome and the native population, and with the impact of the arrival of Rome being minimised (1990a, 8). In this regard Millett suggested this was more of a 'change at the top' adding a new Roman upper tier, combined with a 'light touch' manifestation of imperialism (1990a, 7). Continuing this theme, he believed there was a low level of centralised control from the centre, at least early in the occupation, he saying (1990a, 8): *"The net effect of this was an early Imperial system of loosely decentralised administration which allowed overall control by Rome while leaving the low-level administration in the hands of the traditional aristocracies."*

This certainly set him against the views of the likes of Hingley (1982, 17) who earlier argued in favour of the province of Britain as being an administered economy, he later emphasising its exploitative nature (2005, 49). Millett (1990a, 8) added that he saw Roman Imperialism as an extension of a manifestation of the competitive elite in Rome itself, his referencing of status aligning him with Finley's earlier work regarding the Roman economy (though also incorporating the conception of long-distance trade based on ceramics data), and with Imperial expansion being piecemeal rather than planned.

More recently Erdkamp (2005, 2), taking the Imperial exploitation position, has used the grain market of the Roman Empire as a cipher through which to appreciate the wider Roman economy. It is particularly relevant to this research given Erdkamp's focus on market integration and connecting supply and demand, pertinent to the extractive industries of the South East during the Roman occupation in terms of the access to markets, transport and connectivity so important to their success. He bases his research largely on epigraphic and philological evidence given the lack of material sources to allow reliable statistical analysis. A 'modernist', he believes Roman society was not dissimilar to that of pre-modern Europe (for example using examples from Imperial Russia) and therefore looks for analogies there with the world of Rome. These include the dominance of agriculture, the high cost of transportation and the experience there of market integration. Erdkamp (2005, 14) specifically separates out wealthier farms from those of the peasantry, saying the latter less frequently participated in the capital market and often had the poorest land, this frequently being overworked. This led to a pattern of diminishing returns, as more of the output was consumed by those working the land the harder they had to do so to produce this return (Erdkamp, 2005, 15). This stood in juxtaposition to the larger, wealthier farms which found it much easier to generate a larger surplus. Sticking with the peasant farmers, he adds that given the high instability in prices for their produce, and the back-breaking nature of their daily working lives, they most frequently aimed to produce just enough crops to subsist, with productivity significantly diminishing above this line. This would mitigate against the peasantry producing any significant surplus, and thus often having little to do with the market.

Erdkamp later looks at the evident dichotomy in the ability of the two levels of farming to distribute their surplus (such that it was for the peasantry), with the larger estates far more able to sell their produce in bulk, in so doing defraying their transportation costs (often through middlemen). Meanwhile the peasant farmers, with their far smaller surpluses, were much more likely to sell excess produce within their own communities (2005, 135). He also looks at the relationship between the larger and smaller farms in the context of market integration (2005, 143), illustrating this through the level to which the market could compensate against extreme changes in annual yields. In the pre-modern world the measures to mitigate against such challenges were either storage or the transportation of surpluses to regions suffering shortages, the latter only possible universally if there was a high degree of market integration (contextually relevant to this work given the distances travelled by the extracted materials detailed below in Chapters 3, 4 and 5). He determines that this was actually lacking in the uniform sense, discussing the Mediterranean not being a macro-region economically but a series of bioclimatic micro-regions (Erdkamp, 2005, 146). In this context long-range trade and shipping lanes did exist, especially between the large cities and their hinterlands, but not everywhere was so connected, with the majority of the smaller and medium-sized settlements being far less integrated and supplied from their own hinterlands (with most grain purchases here local in nature). Lavan (2014, 1) adds another factor here in terms of market integration, namely proximity to the coast. He says there was a differential in the experience of such integration between landlocked inland regions and those with a coastline. Using data from the distribution of ceramics as an example, he says (2014, 3): *"...(inland) regions found it more difficult to participate in inter-regional trade (in the context of market integration) because of high transport costs involved in moving goods by river, or especially road, in comparison to those moved by sea."*

Bonifay (2014, 557) expands on this idea, using Roman Africa as an example and basing his conclusions once again on ceramic data. He argues that Mediterranean patterns of consumption were only visible along the strip of coastline adjoining the sea, adding that in inland regions local production and markets substituted for more recognisable Mediterranean products. Again the differential cost of transport is the key factor, and this is a major element defining the provincial (and indeed Imperial) economy. In this regard Hingley (2005, 106) argued that the Roman market system certainly seemed to be at its most integrated in the areas which had maritime access.

Returning to Erdkamp, another dichotomy he identifies is again geographical, but this time on a macro-scale. Given the often-present lack of market integration, he argues (2005, 279) that the west and east of the Empire had different means of dealing with occasional harvest shocks. Where such integration was lacking, he says that in the west the shortfall was often taken up by local aristocrats keen to maintain the regional societal *status quo*. In the east however, permanent institutions were in place to deal with such crises, perhaps reflecting their longer exposure to post LIA classical culture.

Finally, as one clear example of substantial market integration, he cites the grain supply to Rome, where he estimates 30-40% of the Sicilian and Egyptian grain was taken as taxes in kind to feed the citizens of the Imperial capital (2005, 222, see Section 2.2.2 for detail on methods of Roman taxation). In terms of how this was transported, his views have been informed by those of both Finley (1999, 128 [original date of publication 1973]) and Rickman (1980, 4). The former argued that much of the grain importation from Sicily and Egypt to Rome was under Government control, with private enterprise being less important. In this context Russell (2013a, 353) later suggested that, in the context of the improved spatial connectivity that was the consequence of improved infrastructure as the Empire invested in communication, state orchestrated redistributive mechanisms played by far the biggest role in the distribution patterns for goods. Rickman (1980, 4) however argued that much of the grain importation to Rome was through the state relying on grain dealers, shippers and warehouse owners, with private enterprise being more important than previously thought. Erdkamp takes a middle ground in this debate between Finley and Rickman, saying that while the private entrepreneurs were vital to facilitate the grain trade in the integrated market, the state closely supervised this through Government-awarded contracts and incentives (this is analogously again very useful when I later consider the occupation-period extractive industries of the South East in Chapters 3 and 4).

Bringing the discussion of the nature of Roman economy up to date, and taking a more 'primitivist' position, Bang (2008, 12) recently challenged its similarity to that of early modern Europe, saying it was more akin to those of large, tributary or pre-colonial Empires such as Mughal India, the Ottoman Empire and the Ming/ Ch'ing dynasties in China. In this context he says it had features that a modern economist would recognise as manifestations of a Smithsonian market economy, but underwritten by the structural weaknesses (from his perspective) inherent in the nature of Empire, for example exploitation, patronage and predation (2008, 204). Gardner (2013, 6) broadly agrees with such bi-polar manifestations of the Imperial economic experience. On the positive side he talks of improved spatial connectivity (in the context of market integration, based on modern interpretations of globalisation), the most obvious examples being the extensive Roman road network throughout the Empire and improved maritime trade. On the negative side he talks of commodification and alienation, these in the context of a post-colonial or even Marxist interpretations of the Empire. The most extreme example of both commodification and alienation would of course be slavery, with Mattingly (2006, 294) describing a thriving provincial slave trade in Britain. Interestingly, in his critique of Erdkamp's work on the Roman grain market, Lemak's (2006) only criticism is that such slavery is the only aspect of the Roman economy (and a major differentiator when compared to western pre-modern economies) never properly addressed. Temin (2012, 121), in his appreciation of the Roman market economy, himself says that while an appreciation of modern economics can be applied to economies of the past (including that of Rome), the issue of slavery requires special consideration.

Despite such misgivings however, and taking all of the above into consideration, I believe that Erdkamp still presents a model (with his analogies with pre-modern western economies) that is useful when studying the extractive industries of Kent and the South East during the Roman occupation, and the society which facilitated them. As mentioned above this is because his focus on market integration and connecting supply and demand provides insight into the extractive industries of the region (especially with regard to access to markets, transport and connectivity). In particular, his emphasis on the high cost of transport illustrates the quantity (see comment above about defraying costs by transporting in bulk) and value of the materials being extracted and transported by the *metalla* here.

2.2.2 The Roman Imperial Economy: Making the Province Pay

Roman provinces were always challenging to finance, though clearly simple subsistence was not actually their purpose, with each Imperial Procurator under great pressure to ensure it also contributed substantially to the coffers of the Imperial *fiscus*. It had after all to be seen to be *pretium victoria* (worth the conquest). Mattingly (2006, 491), in his stark assessment of the experience of Britain under the occupation of Rome, is very clear about the economic drivers of the Roman state in this regard, saying: *"The economy of Britain was profoundly affected by the desire of the Roman State to extract resources from the Province and this was a constant of Roman Imperialism."*

Hingley (2005, 49) similarly emphasises the exploitative nature of the Empire, directly linking Imperial expansion with the exploitation of human labour in the form of slavery, both he and Mattingly standing in stark contrast in their interpretation of the experience of the provinces to Roman rule when set against that of Millett, above. In the context of this ongoing debate between those with a positive view of Romanization and the post-Romanization focus on experience, identity and community, others more recently have not taken such a bleak view as Mattingly and Hingley. For example Millett himself (2015, 558) has recently highlighted the strong appetite for the symbols of the new culture as shown by data from the excavations along the Roman road from Brough-on-Humber to York as it traverses Hayton and Shiptonthorpe in the East Riding of Yorkshire. Meanwhile, Willis (2008) in his earlier review of Mattingly's 2006 work speaks of continuities enduring from the LIA, being remade in spite of Roman refashioning and impositions. Nevertheless, the Mattingly/Hingley exploitative view, highlighting the not-to-be-doubted material demands necessitated by the need to maintain the empire, is useful here given the nature of the *metalla* considered in Chapters 3 and 4, and is therefore a thread followed below.

In terms of the Imperial economy, and continuing to use Britain as an example, in the first instance such exploitation in the broadest sense was quickly evident during and immediately after the campaigns of conquest. In that regard the brutal exploitation of the spoils of war by Rome to offset the expensive costs of conquest were among the first signs of the presence of the Empire. Here, Mattingly (2006, 494) identifies private possessions in the form of portable wealth as being primary targets of the armies of conquest, with the redistribution of the territories of the landed elites being another. These approaches were of course well-rehearsed in Julius Caesar's campaigns in neighbouring Gaul in the 50s BC. Once the conquest was complete however, and a territory settled into the Empire, the primary demands on the Imperial economy would be as follows (Mattingly, 2006, 493):

- The army, in terms of pay, bonuses, discharge bounties, materials, equipment and supplies (noting in Britain's case the exponentially large scale of this military presence given the province was a border territory, see Section 2.6.1). To give context here, the total costs of the salary and discharge bounties for the army in the 2nd century AD was 150 million *denarii* annually (Mattingly, 2006, 493).
- The provincial Government infrastructure, noting for example that the annual salary for the Governor alone in Britain was up to 200,000 *denarii* (Mattingly, 2006, 493).
- Transport costs, often facilitated by the state (see above discussion referencing Finley, Rickman and Erdkamp with regard to the Roman grain supply in Section 2.2.1).
- Capital investment, including the costs of running Imperial properties and public lands.
- Diplomatic subsidies, always a consideration in Britain given the unconquered far north.

These demands were clearly huge in terms of the burden they placed on the province, especially early in the occupation. In this regard Mattingly (2006, 493) estimates that the overall cost of running the province of Britain at this time would have been *'some tens of millions of denarii'*, and that outside of the spikes in military campaigning activity such as the Agricolan Campaigns of the late 1st century AD and the Severan campaigns of the early 3rd century AD (Elliott, 2016, 129 and 153).

If we continue on this hard line view of the exploitation of the provinces by the Roman state and Imperial household, then the various options available to the state through the Imperial economy to ensure the province paid its way included:

- The property of the landed classes, always at risk of being appropriated and redistributed by the state when an opportunity presented itself, their individual fortunes being equally at risk of a sudden change in providence

Figure 7: Evidence of the Imperial economy. Lead seal from the villa estate at Teston in the upper Medway Valley, associated with the nearby ragstone quarry. Simon Elliott.

for the owner (with confiscated estates often being received by the Emperor in the form of a legacy after such an event, see Section 2.2.4).

- Land controlled directly by the state, in its most extreme in the form of Imperial Estates. These are considered in depth below at Section 2.2.4 given their importance to this review of the extractive industries in Kent and the South East during the occupation.
- The exploitation of the wider population through regular taxation (see below), tribute demands, liturgies and labour requirements, military recruitment and slavery.
- With regard to rural populations, the use of exploitative tools such as rents, dues, price fixing, requisitioning (the appearance of lead seals in the archaeological record such as that seen in Figure 7 can often indicate this) and - once again - tax.
- The exploitation of natural resources, for example the *metalla* covered by this research. Again this could come in a number of forms, up to and including being state run, in the most extreme form as Imperial Estates as set out above (and noting the wide variety of Imperial Estate types, not just those associated with industry relevant to this research).
- Profit derived from the existence of harbours, markets (detailed below in the discussion regarding the provincial economy) and trade (which Temin says was specifically stimulated by the *Pax Romana,* 2012, 2), again through taxation (for example indirect taxation through harbour dues) and surcharges. Also included here are customs charges at the borders of the Empire, either from within-to-without the Empire (or vice versa) or inter-provincial in nature. The *portoria* standard dues when traversing provincial boundaries were paid near to such borders, though some goods such as those supplied under army contracts were exempt.
- The system of military supply, incorporating a number of aspects of the above. In Britain demand from this source had a marked effect on the economy of the province, involving peaks and troughs in the required output and often long distances for such goods to travel (for example from the Wealden iron working sites to the northern border, Cleere and Crossley, 1995, 83).

A key theme running through all of the above means of financially exploiting a province for the betterment of the Empire was taxation. For much of the Principate, this was based on a periodical census listing the resources of a given province which allowed two direct taxes to be levied. These were the *tributum soli* (based on land) and the *tributum capitis* (based on capitation). Such *tributa* direct taxes were ultimately paid to the Procurator, with Mattingly (2006, 496) saying that tax-farmers (working under state-contracts), Imperial fiscal officials and local authorities often acted as middle-men to ensure the smooth running of the system. The latter would have included local elite groups who, through a series of complex interactions with the Imperial aristocracy, had gradually become incorporated into the Imperial economic system (Hingley, 2005, 49). Meanwhile, *vectigalia* indirect taxes (such as the harbour dues referenced above) were collected by officials such as *publicani* contractors. While such methods of taxation were the rule, there was clearly regional and inter-provincial variation across the geography of the Empire and the chronology of the Principate, with individuals and communities often seeking exceptions in some shape or form. Further, as detailed above in Section 2.2.1, given the lack of market integration across much of the Empire, it was often easier to collect taxes in kind, for example in the context of the grain supply to Rome from Sicily and Egypt (see references to Finley, Rickman and Erdkamp above).

2.2.3 The Roman Provincial Economy

Running parallel to the Imperial economy was the provincial economy, as detailed above featuring regional markets, free trade and localised patterns of consumption (Mattingly, 2006, 496) and often difficult to distinguish from its Imperial fellow economy. Its nature was heavily influenced by the pre-conquest economy of a given province, especially at the beginning of a territory being incorporated into the Empire when it adapted what was already there (Millett, 1990a, 66).Thus in Britain it has been argued that the market economy in the South East was more sophisticated than elsewhere in the province given the existing links with northern Gaul (Blanning, 2014, 484).

Continuing the use of Britain as an example, from the point of conquest and the growth of the province it seems that the provincial economy here grew extensively, particularly during the years of the Principate, such that it catered for the top-down needs of the state (and so feeding into the Imperial economy) but also for the downwards-up demand from a new consumer class feeding on new ideas and innovations. With regard to the latter, Hingley (2005, 108) highlights the subordinate cultures of craftsmen, merchants and freedmen who, seeking prosperity and security (in a way that had not been possible in the small scale societies that existed in their given regions prior to incorporation into the Roman Empire), drove a marked increase in trade and industry. For a specific example highlighting data to support this view see the Millett reference above to research at Hayton and Shiptonthorpe (2015, 558), while on the broader subject of increased provincial market demand Mattingly (2006, 497) earlier said:

> *"If measured in simple terms across the period 50-350 (AD), there is plenty of evidence for the evolution of urban markets and the integration of rural territories with them, of an increase in coin use, of expanded manufacturing activity and increased consumption of a wide range of goods across a broad spectrum of sites (military, urban and rural)."*

Specifically on the issue of coinage, he adds (2006, 497) that this was initially part of the Imperial economy given the need to pay the military and administration, with Hingley (2005, 108) earlier arguing that the movements of groups including soldiers created a system of contacts that was quantitatively significant (thus proving highly influential in the provincial economy within which they resided). Nevertheless such coin use only slowly became part of the provincial economy, in a British context not reaching a level with that used in the Imperial economy until the 2nd century AD. The availability of coinage was then disrupted during the 'Crisis of the 3rd Century' when supply was dislocated and poor quality forgeries entered circulation to compensate.

The key component of the provincial market economy were the markets themselves (mostly located at urban centres) which would have been officially sanctioned by the state (showing again the ever present proximity of the Imperial economy). Such markets, which Temin (2012, 6) says knit the Roman economy together such that it could actually be termed a market economy (though again not in the manner of a modern conception of a market economy), acted as emporia for the distribution of manufactured products, many innovations of the occupation. A key issue here is the inability to identify whether demand for many of these goods would have derived from local communities, or from manifestations of the Imperial economy such as the military and administration.

To conclude this sub-section, the experience of the provincial economy was often bi-polar in terms of market integration as detailed in Section 2.2.1, based on the geography of the territory, with Hingley (2005, 106), Lavan (2014, 1) and Bonifay (2014, 557) outlining the importance of being near the coast or significant riverine systems to get the full benefit of participating in an integrated market. Evans (2014, 438) too develops this theme in his own consideration of the Roman economy, yet again using ceramic data to explain that the economy of a given area (in the case of his research Britain) always featured a balance between local, regional and inter-regional trade and economic engagement, with geography (through relative transport costs) being a constant factor affecting the balance.

2.2.4 Imperial Estates

Mattingly (2006, 494) says that the exploitation of natural resources (as part of the Imperial economy in a province) was a major area of state control. Referencing particularly the *metalla*, this would have been in the form of direct control, at its most extreme as an Imperial Estate. It was through the Imperial Estate model that the largest percentage of derived revenue could flow most ergonomically to the Imperial coffers, with the least finding its way into the regional market economy. In that sense it is one of the reasons archaeologists and historians have often cited a lack of display of disposable wealth at a local level, for example in the form of villa estates, as an indicator of the presence of an Imperial Estate. Similarly, Mattingly (2006, 495) says there is little evidence that individual *civitates* benefited from their proximity to such resources, another indicator of a significant state presence in the form of such unusual land use patterns.

Given the importance of the Imperial Estate model to this research, with its focus on the *metalla* of Kent and the South East during the occupation, here it is examined in detail. This will include a definition of what an Imperial Estate was and their types, their management, the evidence used by archaeologists and historians to identify them (especially when lacking written and epigraphic evidence), the rebuttals to such identification to provide balance, and finally a summary to allow an Imperial Estate template to be developed which can be utilised when considering the nature of the occupation-period extractive industries of Britain.

Crawford (1976, 36), in her detailed study of agricultural Imperial Estates, emphasised that such geographic economic entities were the personally owned landed wealth of the Roman Emperors, the same being true of their industrial Imperial Estate counterparts. She said (1976, 36) that the latter included the large scale *metalla*, brickworks and salt pans, though see below with regard to the latter.

Such Imperial Estates came into the possession of the Emperor in a variety of ways. Mattingly (2006, 455) says that Imperially owned (and indeed other state-owned) land was often the result of confiscation at the point of conquest from regional elites (Salway, 1981, 104 cited the royal estates of the Catuvellauni becoming the private possessions of the Emperor after the AD 43 invasion) or later in terms of failed usurpation attempts and other revolts. They could also be the result of inheritances from the provincial elite, the Imperial household being one of the largest provincial landholders. Crawford (1976, 38) gives an excellent example of the latter, based on written records, in the form of the Emperor Hadrian's sister-in-law Matilda bequeathing extensive agricultural estates around Sitifis in Mauretania to the Imperial household. Meanwhile, as detailed above, many commentators (Crawford, 1976, 36 as an example) have long argued that the *metalla* and other significant industrial enterprises naturally fell under the immediate aegis of the Emperor as Imperial Estates.

In terms of how Imperial Estates were managed, below the administrative level of the Procurator's office, options could have included the military where appropriate (often in terms of opening up the opportunity in the case of the *metalla*, Mattingly, 2006, 507, see above and Section 2.6.3), *vilici* (bailiffs, singular *vilicus*), chief tenants (in the form of a head lease *conductores* for example) or other tenurial arrangements, and to confuse matters any combination of such (this also applying to non-Imperial state-owned land, adding an extra layer of confusion). Mattingly (2006, 455) is rather blunt in his interpretation of the role of such Estate managers, saying in a British context: *"Their function was to cream off surpluses from the British occupiers of such lands."*

A good example of one such *vilicus* comes from the exceptionally large eastern/ coastal Wealden iron working site at Beauport Park, with a reference on the bath house stonework entrance to such an official who ran the site (Brodribb *et al*, 1988, 261, see Section 3.1.3).

I now turn to the evidence used by archaeologists to identify Imperial estates. This includes:

- The written and epigraphic record. Crawford (1976, 37) for example highlights papyrus *P. Bouriant* 42 from Egypt dated AD 167 which features an administrative survey of 3,032 ha of land of which 39% were agricultural Imperial Estates, while the Beauport Park *vilicus* above provides another. Such evidence is clearly the firmest method of identifying such geographic economic entities, though see discussion at Section 6.1.3 regarding the latter example. Other epigraphic examples are more contentious, for example that from the villa estate at Combe Down near Bath (RIB 179) referencing an Imperial freedman named Naevius who carried out reconstruction work for the Procurator, which it has been argued might indicate a possible Imperial Estate (Crawford, 1976, 36, and Mattingly, 2006, 399) related to quarrying. One can perhaps add here the large quantities of *Classis Britannica* stamped tile often associated with a State presence at some of the key iron working sites in the eastern/ coastal region of the Weald, including Beauport Park (Brodribb *et al*, 1988, 275, see Section 6.1.3 and Section 6.1.4 for full discussion below).
- The lack of villa estates in a landscape featuring other activity, for example large-scale industry (Salway, 1970, 10). Cleere and Crossley (1995, 68) have for example highlighted this in the context of the eastern/ coastal iron working sites in the occupation-period Weald where such elite settlement is singularly lacking, though this is not true elsewhere (see below).
- Sticking with settlement, other atypical rural settlement patterns (Mattingly, 2006, 455). A good example of one such is the interpretation of the stone tower (building R1, Potter, 1996, 678) and settlement at Stonea in the Fens, Cambridgeshire as a regional administrative centre (Potter and Jackson, 1982, 118) associated with an Imperial Estate. Here some (for example Potter, 1996, 688, and Malim, 2005, 126) have argued that this central multiple-story stone building with its surrounding gridded street pattern (for largely non-stone built structures) is a better fit for the model of state-intervention than private enterprise, this view first presented by Salway (1970, 10). Artefacts recovered here have also been used to interpret the site's official function, for example writing tablets, a stylus and evidence of luxury food stuffs, as has the difficulty in supplying the stone along regional waterways to build the central structure on this upland 'island' site amid the peat fens. Further, Potter (1996, 685) talks of several pointers towards a military presence at the Stonea site, for example military metalwork (though note Taylor's contention regarding a military interpretation here, 2000, 657). In this regard, Malim (2005, 126) specifically references the use of the military in facilitating the running of agricultural Imperial Estates such as that argued for at Stonea.
- Unusual archaeological data, for example ceramic evidence supporting the significant expansion of the farming activity around Stonea in the early 2nd century AD (Hartley and Hartley, 1970) which has been used to date the initiation of the Imperial Estate there (if such it was) to the visit of Hadrian to Britain in AD 122 (Salway, 1981, 190, and Potter, 1996, 678). Malim (2005, 129) later speculated, based on data from the original investigations, that the land itself may actually have come under Imperial control after the Boudiccan Revolt of AD 60/ 61, later being fully exploited from the time of Hadrian's visit and reaching its full potential by the time of Antoninus Pius.
- Unusual and unattributed developments in transport networks. Sticking with the Fens as an example, canals provide one example where Fincham (2002, 10) highlights the arguments in favour and against interpreting the Car Dyke as an occupation-period transport route. New and often raised roadways (the Fens Causeway being the best example, though noting this dates to the mid-later 1st century AD) are another in this region, both transport modes being interpreted as evidence of the Stonea location being an Imperial Estate (Malim, 2005, 126), requiring as they do specialist surveying and engineering skills in their construction (if indeed the Car Dyke was a transport route).
- Unusual developments in land usage. Again using Stonea as an example, this includes significant land reclamation with drainage (which again would have required specialist skills in terms of surveying and engineering), and secondary land division (Malim, 2005, 126), though see rebuttals by Mattingly, Millet and Taylor below.
- Demand, in the case of Stonea from the Nene Valley to the west of modern Peterborough given the rapid

population expansion there from the 2nd century AD and for the military in the north of the province, easily supplied from this location given the easy maritime access, a pull factor for the Imperial Estate's produce if indeed it was one (Malim, 2005, 128). One can perhaps add here demand for building stone in Roman London for the ragstone quarries of the upper Medway Valley through to the mid-3rd century AD (Elliott, 2016, 102), this being discussed in depth in Sections 4.5 and 5.1.4.
- Using Stonea one final time as an example, other industrial activity associated with Imperial Estates. Here this is salt manufacturing as referenced by Potter (1996, 688), which Salway (1970, 10) originally argued was an Imperial monopoly.

I now turn to the rebuttals of these various means of identifying Imperial Estates given the fact that outside the very hard examples of written and epigraphical data, the various models and methodologies of identifying these geographic economic entities are a matter of judgment by the individual archaeologist and historian, even when backed by a substantial body of archaeological data.

In the first instance Mattingly (2006, 371) does not see Imperial Estates as being so black and white in terms of their identification. Noting, as detailed above, that an absence of villa estates has often been used to identify an area as a potential Imperial Estate, he believes there are no reasonable arguments for believing Imperial Estates lacked villas, given (he argues) such properties were frequently contracted out to private individuals and run for profit. By way of example, he continues (2006, 371) that in North Africa large Imperial Estates (at the heart of the agricultural zone, such as those detailed in papyrus P. Bouriant 42 above) were dotted with farms, villages and villas. He concludes (2006, 372) that the more probable candidates for Imperial Estates might in actual fact be found in areas featuring higher villa estate densities, where rural identity and culture stood in contrast to the more average development. This is an important consideration in Chapter 4 and the discussion in Chapter 5 regarding the ragstone quarrying industry in the upper Medway Valley in particular, given villas are embedded within that occupation-period industrial landscape (though noting Mattingly's examples above are largely agricultural Imperial Estates). Importantly, he then identifies the variety of candidates who might occupy such villas (again very relevant to this work, in the context of those embedded in industrial Imperial Estates). These include representatives of the Roman Emperors and the State, absentee landowners, the army (including the regional fleets such as the *Classis Britannica*, Elliott, 2016, 108), members of the *civitates*, settler groups (for example discharged soldiers with investable capital), private individuals (from the home province and abroad), religious bodies and entrepreneurial bodies.

Mattingly also adds, in the context of atypical rural settlement patterns, that in the absence of hard epigraphic evidence it is difficult to detect large agricultural Imperial Estates on the ground given they possibly resemble other types of agricultural settlement (2006, 455). Meanwhile Millett (1990a, 120) has also argued that there are a number of problems with the identification of land as an Imperial Estate. These include his belief that there is no reason to suppose that land owned by the Emperor is archaeologically distinguishable today from other types of land use. Additionally, with regard to agricultural Imperial Estates, he says that if they were created from virgin territory, there would be more signs of a deliberately even distribution of land, for example by centuriation, and this is often lacking (he uses the Fens as his example). Meanwhile, regarding the issue of unusual land use and again using the Fens as his example, he says there is no reason to believe that local populations would not have had the appropriate levels of societal organisation needed to reclaim and develop their own land. Taylor (2000, 652 and 653) also contends the unusual land use patterns claimed for the Stonea site, and the interpretations of the principal stone-built structure, saying that it had suffered such a degree of destruction that any attempt at reconstruction was highly speculative. He (2000, 654) continues his questioning of the Imperial Estate interpretation of the Stonea site by arguing that the canals and roads of the Fens were not part of an Imperial project but the result of localised improvements to an existing network, adding (2000, 655) that the Car Dyke may actually have been a catch water drain.

Finally in these considerations, Millett (1990a, 121) debunks Salway's earlier assertion (1970, 10) that salt production was an Imperial monopoly in the province, basing this on coastal salt production sites being located and excavated at numerous other locations since the original excavations at Stonea, thus undermining this argument in favour of the Fens (in this example) being an Imperial Estate. Taylor (2000, 655) is also critical of the use of salt production to interpret the Fens as an Imperial Estate, he saying among other things that given salt production in the few such sites excavated in the region ranged from the LIA to the 4th century AD, this mitigated against them being used as evidence of a Hadrianic Imperial Estate.

I conclude this sub-section by reflecting on the above discussion regarding the nature of Roman Imperial Estates as part of the Imperial Economy, aiming to determine a usable model going forward in this research. Clearly, outside of the use of definitive written and epigraphic data to support such a geographic economic entity being identified as an Imperial Estate, classifying one as such is problematic. This is further complicated by the fact that the running of Imperial Estates themselves were often sub-contracted out, as were other state-run entities within the Imperial economy, obscuring attempts at their determination. Such identification must therefore be a question of individual consideration by the researcher, based on the available data. The issue can be helped however by utilising a general though heavily caveated model (noting the reservations of Mattingly, Millett, Taylor and others above) which can be utilised as a concluding check list after considering the data for each candidate in detail to provide focus. This will help to present a viable if imperfect final template after all of the data has been considered, to be used when concluding the relevant sections in the discussion at Chapter 5 regarding the state presence in the *metalla* of Kent and the South East during the occupation (these at Sections 5.1.3 and 5.1.4). The specific features I identify in this model are:

- A lack of settlement other than the living quarters of the workforce, whether villa estates, non-villa settlements/native settlements or small towns, where one might expect to find them. In the case of such settlements, common sense dictates this is more applicable to industrial Imperial Estates than to agricultural Imperial Estates, where Crawford (1976, 37) made the case that the latter are not always easy to interpret and where Mattingly (2006, 371) references examples of such settlement in North African agricultural Imperial Estates (see above).
- 'Federal' level investment in unusual transport networks. While I note above the caveated nature of the features in this list, and especially in this case the examples noted above regarding Stonea (Taylor, 2000, 655), I still believe it valid to include the feature here. This is because I feel that, provided unusual transport networks are considered holistically alongside the other features listed, they do add some value, particularly with regard to two examples important to this research. The first is the Wealden road from Rochester through modern Maidstone which terminates just north of Hastings amid the occupation-period industrial-scale iron working sites located there (Staveley, 2013, see section 3.3 below for full detail). The second is the possible canalisation of the occupation-period upper Medway Valley, discussed in Section 4.5.2 in the context of facilitating the ragstone quarrying industry there (Elliott, 2016, 105).
- Unusual land use patterns, where again I note the above caveated nature of the features in this list (for example Taylor, 2000, 652 and 653 regarding this specific feature), but still believe this to be a valid inclusion. Once more this view is in the context of the research presented in the work, given the insight it provides into the differing natures of industrial activity in the Weald and upper Medway Valley.
- An association with other industry (for example see Section 3.6 below regarding tile and brick manufacturing alongside the iron industry in the Weald).
- The presence of the Roman military in an unlikely setting, Potter (1996, 685) for example highlighting the presence of military metalwork at Stonea as potential evidence for a state presence there, while evidence for a *Classis Britannica* presence in the Weald is considered in the discussions at Section 5.1.3.

2.3 Regional Background Overview

The geographical region now called Kent was a vital component of the Imperial experience in Britain, featuring at the very beginning of the occupation (almost certainly being the point of entry for the Claudian conquest, see Section 2.2.1) and indeed at the very end (when Richborough was still an active Imperial outpost). It is of vital importance to this work given the location of much of the Wealden iron industry and all of the upper Medway Valley ragstone quarrying industry there, both the subjects of this research. To that end there are five broad themes concerning the Imperial experience in Kent which I set out here given they provide a useful template for the later regional research in Chapters 3 and 4, and discussions in Chapter 5. Specifically, these themes are the Roman occupation of Kent, settlement in Kent during the occupation, the population of Roman Kent, the economy of occupied Kent (both in terms of geography and chronology), and the end of the occupation in Kent.

2.3.1 The Roman Occupation of Kent

Kent was known to ancient sources as Cantion before the occupation, being populated by the Cantiaci (Jones and Mattingly, 1990, 44). It was bordered in the immediate pre-Roman invasion period by the Trinovantes to the north, the Catuvellauni to the north west and the Atrebates and Regnii to the west. Caesar (The Conquest of Gaul, V.135) describes Britain at the time of his 55 BC sortie as being densely populated and heavily cultivated, and it is likely he was referring here to Kent given his probable landing place on the eastern coast of the county, both in that year and also in his later landing in 54 BC (Millett, 2007, 115). Given its geographical position, Kent has often been at the forefront of cultural change in Britain originating from the Continent (Wenban-Smith, 31, 2007). This noticeably accelerated in the LIA prior to the definitive Roman invasion of Claudius (Emperor AD 41 – AD 54) in AD 43, as the economic and later political reach of Rome spread aggressively northwards from the Mediterranean (Haselgrove, 1987, 116, Cunliffe, 1988a, 108, and Cunliffe, 2013, 360). As Champion (2007, 129) details:

> *"The final two centuries of the Iron Age saw major changes in the nature of society in South Eastern England, including imports from the classical world (such as wine amphora, Italian bronze vessels and South Gaulish samian ware), the emergence of new types of settlement, the adoption of coinage, the appearance of a new mode of burial, and new styles and techniques for the production of pottery."*

Such dramatic economic and social change in the LIA manifested itself in other ways too. For example Haselgrove (1997, 71) uses archaeological data from Iron Age brooch finds across the county to argue for the appearance of new fashions and means of adorning the body, while Hill (1997, 106) interprets data from toilet instruments and vessels for washing as indicating a new focus on bodily hygiene among some elites. Champion (2007, 131) believes that together these latter two developments show that a major change was taking place in the nature of human relationships. Meanwhile, Blockley and others (1995, 42) also argue that LIA data from the excavations at the Marlowe car park site in Canterbury, featuring a potsherd with graffiti and the use of inscribed coins, indicate a new (though limited) level of literacy. Finally, Champion (2007, 131) cites architecture as an example of LIA cultural change in Kent, specifically the increased appearance of rectangular plan buildings and fired-clay building materials.

Political change is also evident in Kent in the LIA. A key factor here would have been the economic impact of the arrival of the kinds of elite goods described above, and seen in quantity in the recent excavations at the likely port of Folkestone, East Wear Bay (Parfitt, 2013, 27). Further, direct interaction between the Kentish elites and events on the Continent would also have played a major role. In this regard, Caesar (The Conquest of Gaul, V.119) is explicit in saying Britons regularly fought against his forces during the conquest of Gaul as allies of the Gallic tribes, he also explaining that Gallic King Diviciacus of the Suessiones (located in the Aisne Valley) claimed dominion over some of the tribes of Southern Britain (The Conquest of Gaul, II.58). Archaeological data to support these LIA political links between Kent and the Continent of the Caesarean conquest comes in the form of the proliferation in the county of Gallo-Belgic E coins minted in Gaul to fund the conflict with the Romans (Cunliffe, 2013, 326), illustrating the shared practices and kinships which characterised these transmarche communities. Some final insight into the political situation in the LIA in Kent comes once again from Caesar, who explains (The Conquest of Gaul, V.139) that at the time of his 54 BC expedition there were four Kings in Cantion, namely Cingetorix, Carvilius, Taximagulus and Segovax.

Caesar seems to have secured tribute from British tribes that was potentially enduring through the later first century BC. Some have argued that the process by which the more prosperous regions of pre-conquest Britain (including Kent) ultimately fell under the political control of Rome was actually more insidious and protracted than simply through post-Gallic conquest cultural contact followed by the 1st century Claudian invasion. Manley (2002, 143) for example says:

> *"An alternative understanding to the events of AD 43 is to consider the events of that year as the culmination of a process of the preceding century, which effectively had subjugated at least the South East of Britain and brought that area within the complete control of the Roman Empire some time before AD 43."*

Whatever the truth here, the Claudian landing was clearly a major watershed. Given its proximity to the Continent, and noting arguments in favour of other potential landing places such as Chichester (Manley, 2002, 131), Kent is still widely argued to have been the landing place for this crucial event (Grainge, 2005, 117).

The most likely landing areas would have been the coast of eastern Kent, with the then shelter of the Wantsum Channel and the broad expanses of beach from Sandwich to Deal being particularly inviting to the invaders, adding to the relatively safe harbourage of Pegwell Bay. With regard to the Claudian invasion Moody (2008, 141) explains that:

> *"Given the size of the operation, it is unlikely that a single location can be identified for the Roman landings. More probably, the ships landed where they could, in the network of harbours, beaches and trading ports on the east and western sides of the Wantsum with the troops securing themselves, by units, over a wide area."*

Wherever the specific landing places were in eastern Kent for this invasion, the locale for the event was later commemorated by the building in the reign of Domitian of a monumental arch at Richborough which stood from the later 1st century through to the late 3rd century, latterly being used as a signal station (Strong, 1968, 72).

Kent was known after the AD 43 invasion as Cantium (featuring the *civitas Cantiacorum*), becoming part of the original province of *Britannia* from the mid-1st century AD through to the early 3rd century AD when it became part of *Britannia Superior* following the reforms of Septimius Severus and his son Caracalla (Mattingly, 2006, 229). Following Diocletian's later reformation, Kent sat within the new province of *Maxima Caesariensis,* one of four (or five) provinces in the new *diocese* of Britain (White, 2007, 51).

2.3.2 Settlement in Occupied Kent

Kent rapidly established itself as a key region within the province, with 12 of the 21 1st century AD villas in the South East being located in the county (Allen, 2013). It sat firmly within the more settled south and east of occupied Britain which thrived as part of the Imperial project, unlike the upland north and far west which Mattingly (2006, 149) argues was economically dominated by the large military presence there needed to maintain the extreme north western borders of the Roman Empire. This division created a geographically bipolar experience for those living in Britain during the occupation, having major implications for Kent given that the principal gateway through which the military presence symbolically entered and left the islands sat within the county at Richborough (Millett, 2007, 143). Newly arriving military units will have been greeted with the striking sight of the presumed gilded equestrian statue that sat astride the Richborough arch (Strong, 1968, 45), whether they landed at Richborough or sailed on through the Wantsum to London.

This concept of the county, and specifically its east in the context of Richborough, fulfilling the role of the province's principal gateway is particularly important and allows us to begin to differentiate Kent from the rest of the south and east. There is clear evidence of this differentiation well before the occupation, for example Ashbee (2005, 118) arguing that there was continual cultural exchange between the county and the Continent given their close proximity, ahead of much of the rest of the region, from at least the Mesolithic period onwards. This predated Britain becoming isolated from the rest of the Continent by rising sea levels, and then continued afterwards as a matter of course. In this regard Yates (2004, 13) details the disproportionate increase in metalwork in the county in the Bronze Age as Kent benefited from its proximity to the Continent, while above in Section 2.3.1 I detail the extensive economic, social and political changes that occurred in Kent in the LIA. This differentiation continued into the occupation, with the region certainly being a place of difference in Roman Britain given the continuing proximity of links with northern Gaul and the Rhineland.

Other examples of the county's occupation-period differentiation, in addition to it being the Imperial Gateway, are considered in much greater detail below in the two regional studies but include the unusually large number of public buildings in Canterbury compared to many other *civitas* regional capitals (Millett, 2007, 169), and the extensive extractive and manufacturing industries that thrived in Kent until at least the middle of the 3rd century AD. Kent's regional uniqueness continued well after the occupation, exampled by data from grave goods and burial practices which show a strong Frankish influence in the Saxon county (Welch, 2007, 220).

Moving on to the chronology of settlement in occupied Kent, Blanning (2014, 480) says that as evidenced by villa estates there is a clearly noticeable pattern. This began with an increase in rural settlement in the LIA which accelerated rapidly with the Roman occupation and the establishment of villa estates, particularly towards the end of the 1st century AD. This is illustrated by the fine example depicated at Folkestone in Figures 8 and 9. There is then a noticeable decrease in new sites in the 2nd century AD, and by the latter part of this century many of the original Roman sites were falling out of use. This decrease in occupied sites continued into the 3rd century AD, with by the mid-4th century AD only a quarter of original foundings still being in use, often in different ways to their original function as an elite residence. A good example in this regard is the temple at the villa site in East Farleigh, part of which became a kitchen with the installation of ovens at this time (see Section 5.1.4).

As today, there are strong indicators that settlement in Kent was uneven during the occupation. For example areas such as the North Downs and Weald were far less densly settled than other areas of the south and east (Blanning, 2014, 128, and Millett, 2007, 170). The Roman Rural Settlement Project (a collaboration between English Heritage, The Leverhulme Trust, Cotswold Archaeology, the Archaeology Data Service and the University of Reading) illustrates this, showing that while more of Kent has been excavated than elsewhere in the region (making up 34% of the total), the county only ranks seventh out of eight in terms of the number of Roman sites recorded per hectare, and by some way (see Table 2.1).

In Kent during the occupation the below areas were the most densely settled, tracking the principal road and marine (both coastal and river) transport routes, fertile soils and industrial activity in the county:

- The countryside along the north coast and along Watling Street.
- The strip of land running along the foot of the South slope of the North Downs.
- The rivers valleys of the Stour, Medway and Darent.

Data from a century and a half of excavations in the county show that Kentish villas had a particularly local quality, for example many featuring underground cellars similar to the practice in adjacent parts of the Continent (see discussion below in Section 4.1.4 concerning the Wouldham/ Burham 'mithraeum'). Blanning (2014, 482) adds that villas in the county tended not to feature large *triclinia*, but instead had particularly large *porticus*. Meanwhile Taylor (2011, 183) argues that villas in the region featured bath houses from an earlier date than elsewhere in Britain, as evidenced at the villa site at Teston on the Medway (the first phase here dating to the later 1st century, Elliott, 2013, 40), indicating a strong desire to display *Romanitas* from the very beginning of the occupation as well as reflecting proximity to the Continent. Additionally, the appearance in the occupation of large contemporary burial barrow mounds associated with villa estates also reflects experiences on the Continent, for example those at Flaxweiler in Luxembourg (Krier and Henrich, 2011, 213) and Koninksem in Belgium (Crowley, 2011, 199). Such barrows and other manifestations of mortuary behaviour, as evidenced by the walled cemetery at Barming (Taylor, Jessup and Hawkes, 1932, 145), were built with visibility in mind, either from their associated villas (as in the case of Teston) or from local roadways, often those leading to the villas. The latter is an example of the importance of the visitor experience, which Catling (2013, 33) says was a key factor when villa locations were being selected. Taylor (2011, 184) adds that the vistas from the villas up and down the Kentish river valleys would also have been a key factor in their location in the county, with that at Teston again being an excellent example and a good candidate for a future viewshed analysis.

2.3.3 The Population of Occupied Kent

Determining the exact size of the population of Roman Kent is problematic, especially given the similar but larger scale issues detailed by Millett (1990a, 185) and Mattingly (2011, 219) for establishing the population of the whole of occupied Britain. With regard to the latter, Mattingly usefully sets out in tabular form Millett's figures set against his (for the 2nd and 4th centuries AD), broken down into three broad communities as in Table 2.2.

Based on the archaeological evidence and also analogy from elsewhere in the Empire, together with my own research, I believe Millett's figures to still be the most accurate. The figure of 3,665 million would be up from a maximum of around two million in the LIA, it then falling back to below

Figure 8: One of the finer Kentish Roman villas. An artist's impression of phase two of the villa at East Cliff, Folkestone, late 2nd century AD. 'A Town Unearthed'/ Drew Smith and Mikko Kriek.

Figure 9: One of the finer Kentish Roman villas. Another angle of the artist's impression of phase two of the villa at East Cliff, Folkestone, late 2nd century AD. 'A Town Unearthed'/ Drew Smith and Mikko Kriek.

two million after the occupation, then beginning a steady climb and reaching up to eight million before the Black Death in the 14th century AD (Cunliffe, 2013, 97).

Looking specifically at the Kentish population during the occupation, Goacher (2009, 391) has attempted to determine its size using the methodology utilized by Millett (1990a, 185) in his above national estimate. In that regard she suggests an urban and regular military population for Kent of 18,000, a rural population in agriculturally productive areas of 51,000, a rural population in the less productive areas of 35,000, and a population in marginal areas (for example the Weald) of 6,000. This gives a grand total of 110,000, 3,000 less than the population of modern Maidstone. Interestingly this figure compares less favourably than other regions in the productive south and east of Britain during the Roman period. For example Greene (1986, 124) speculates that that Northamptonshire would have had a population of up to 150,000 in the same period. Such a differential between two modern counties nominally featuring a similar occupation-period regional economy is best explained by the fact that the less productive rural and marginal areas in Northamptonshire, and indeed elsewhere in the region, were better utilised than in Kent.

2.3.4 The Economy of Occupied Kent – Geography

In terms of economic activity in Kent during the occupation, a number of leading commentators have set out their views based on settlement patterns, numismatic data and pottery data that during the Roman period there were three distinct geographic regions in the county. These views are set out below for later exploration in the regional surveys in Chapters 3 and 4 and discussion in Chapter 5.

Table 2.1: Roman Rural Settlement Project – South East

Sites per Hectare	County
1.45	Oxfordshire
1.45	Buckinghamshire
1.41	Surrey
1.21	West and East Sussex
1.2	Hampshire and Isle of Wight
0.85	Berkshire
0.45	Kent
0.4	Greater London

Allen, 2013.

Mattingly (2006, 386, in what he describes as a minimalist reading of the evidence) says that the Weald, rich in iron siderite and timber but thinly settled, was a very distinct entity which could have been under some kind of Imperial control. Meanwhile, again based on a minimalist reading of the evidence, he limits the specific *civitas* of the Cantiaci to north eastern Kent (as opposed to this being the entire area of the modern county, 2006, 386). He indicates this north eastern area was distinguishable from the north western region of the modern county (centred on the river valleys of the Darent and Medway, which featured many of the region's villa estates) where the land may have been leased or sold to private land owners. Meanwhile Moorhead's (2014, 117) analysis of Roman coins in Britain using data from the Portable Antiquities Scheme (PAS) also highlights the same three distinct geographic economic areas, with a comparative lack of coins found in the Weald (though caution should be used here given the likelihood of less detectoring in this area) and dense concentrations in both the eastern region and in the north western river valleys.

Next, Lawson and Killingray (2004, 20) say there were broadly two major centres of pottery production in occupied-Kent which also reflect the above regional economic analysis, these being the Canterbury-wares of the eastern zone and the Thames-wares to the northwest in the Thames/ Medway Estuary. Lyne (1994, 126) says that the latter was particularly successful by the 2nd century AD, dominating the west of the region but also being found in quantity in London and across eastern Surrey where it competed with the western Weald's own Alice Holt pottery industry. Lyne's research (1994, 428) goes on to show that this polarisation of the eastern and north western pottery industries, with comparatively little in the Weald excepting at its western edge, continued well into

Table 2.2: Roman population in Britain

Community Type	Sub Unit	Millett 1990a	Mattingly 2nd C	Mattingly 4th C
Military	Army	10,000 – 20,000	45,000 – 55,000	20,000 – 25,000
	Garrison settlement (vici and similar)	50,000 – 200,000	100,000	50,000
Urban	Major towns	183,971 – 290,057 (both major and small towns	120,000	100,000
	Small towns		25,000	50,000
Rural	Villa dwellers	Mid-point average of 3.3 million, both villa and non-villa.	5,000	60,000
	Non villa settlement		1,700,000	2,215,000
Total		**3,665,000**	**2,000,000**	**2,500,000**

After Mattingly, 2011, 219.

the 4th century AD even taking into account the decline of Thames-ware industry from the 3rd century AD.

Each of these three occupation-period economic zones had differing regional economic allegiances (Moorhead, pers. comm. 26 February 2013), with the Weald (especially the eastern/ coastal area, see Section 3.4 below) looking to the coast from where manufactured materials were exported both around Britain and indeed to the Continent. Mattingly (2006, 386) argues that the Weald remained under some form of Imperial control (see discussion in 5.1.3 regarding this region being part of the Imperial economy, and the case for it being an Imperial Estate or not). Intensive industrial activity here ceased in the mid-3rd century AD, at which time much of the region was abandoned (Cleere and Crossley, 1995, 81). This intense Principate industrial presence, and the subsequent regional abandonment, would help explain the comparative lack of settlement in the Weald itself, and especially the lack of small towns and villas in the Weald in Kent (excepting the small town of Westhawk Farm at its extreme eastern edge) and West Sussex, and the similar lack of such sites in Wealden East Sussex save the possible roadside settlement at the major iron working site at Footlands Farm, recently identified by the Hastings Area Archaeological Research Group (HAARG) through geophysical survey (Cornwell, 2013, 14, also see Sections 3.1.4 and 3.2 for discussion regarding the sites at Garden Hill and Bardown). Indeed the main regional settlements seem to have been on the periphery of the Weald at Barcombe (a villa, with a small settlement now also recorded at nearby Bridge Farm), Buxted, Penhurst, Hassocks and possibly Arlington (Cleere and Crossley, 1995, 58).

This pattern in the Weald, with the majority of settlement as mapped to date being found on the periphery of a major Roman industrial zone, has analogous parallels elsewhere in the Empire, for example the industrial region at Autun in eastern France (Harrington and Welch, 2014, 110). Given the fact that minerals were designated an Imperial monopoly in the Empire (Mattingly, 2009, 296), one can reasonably argue that one of the explanations for this lack of settlement in the Weald was active discouragement by the state to preserve the vital raw materials located there. This would have included not only siderite for the iron industry but also clay for tile manufacturing and trees in the extensive Wealden forests for their wood to support industrial activity, and also for building and ship construction.

Meanwhile, the eastern region was under the control of a separate sphere of economic influence centred on Canterbury and stretching to the Channel coast as far south as Folkestone, while the north west was again distinct from the rest of the county and, as set out above, possibly leased or sold to private landholders. This north western zone may have specifically been in the economic sphere of London throughout the occupation, with Mattingly (2006, 386) also saying it was possible some land in Kent was assigned to London.

These three regional areas of economic activity may actually have had their roots in the LIA based on coin and pottery data (Holman, 2000, 220, and Haselgrove, 2005, 1), especially if one were to use earlier processual models linking *oppida* defended settlements with specific hinterlands. Examples would be that at Boughton Monchelsea with the Weald (as argued by Howell, 2014, 40), those at Canterbury and Bigbury with the east and that at Oldbury with the north west. In this regard Cunliffe (2009, 237) still believes that authority across the south and east of Britain in this pre-Roman period was centred on these *oppida,* from where he argues elites controlled a subsistence economy dominated by grain production.

The stamp of the state may have actually been county-wide and not just restricted to the Weald, this interpretation evidenced by the preponderance of public buildings in Canterbury as mentioned in Section 2.3.2, some being conspicuously very large such as the theatre, and the comparative lack of nearby villas to the *civitas* capital (Taylor, 2011, 183). Andrews (2001, 25) here draws on comparisons with towns in frontier regions where the military did not invest in villa architecture, one hypothesis being that in a Kentish context this was because of the domineering effect of the military in the county. He argues that through much of the occupation Canterbury was more of a centre for local military and state representatives to gather rather than being the more traditional type of *civitas* capital catering for an extensive civilian population. Taking his comparison with frontier regions further, a useful analogy might be the north and west of occupied Britain where the military presence was clearly dominant (Millett, 1990a, 100). Additionally Andrews (2001, 25) believes that Kent's proximity to London added to a sense of militarisation across the modern area of the county given that both the Governor and Procurator resided there, and later the *vicarious.* An example would have been their assignments of troops and state functionaries, some of whom may have lived, at least for some of the time, in Kent.

2.3.5 The Economy of Occupied Kent – Chronology

In terms of chronology, Pollard (1988, 198) uses ceramic data from his extensive review of occupation-period pottery in Kent to argue that like much of southern and eastern Britain the county seems to have flourished for much of the early part of the occupation. Millett (2007, 159) similarly says that the growth in the number of buildings constructed in a 'Roman' style and the spread of the use of similarly 'Roman' objects in everyday life across the county from the late 1st century AD is also a key indicator of economic prosperity. This was not to last throughout the occupation however, with both Blanning (2014, 480) and Mattingly (2006, 388) saying that by the 4th century AD, when other parts of the *diocese* such as the South West were experiencing a late resurgence, this was singularly not evident in Kent (Blanning, 2014, 480). Lyne (1994, 127) adds further evidence here, referencing the decline of the previously resurgent southern market for goods from the Thames-ware pottery industry from the later 3rd century AD.

For the earlier and seemingly economically more successful period, archaeological data at the major harbour settlements which developed at Richborough and Dover (Millett, 2007, 169) indicate that this success was assisted by cross channel trade, with Philp (1982, 178) also believing Lympne may have been such a commercial centre. This trade was driven by two factors:

- The desire on the part of many elites to attain and display the trappings of *Romanitas*, this being magnified by proximity to the Continent as set out above. Pollard (1988, 198) highlights here the import of elite pottery from the Rhineland and Gaul, while Millett (2007, 159) highlights the similar import of objects for use in everyday life from the Continent.
- Also, as detailed above, the disproportionately large military presence in Principate and later Dominate Britain. Moody (2008, 156) specifically sets out his view that intensive demand in the South East region was driven by the military presence in Britain during the occupation, with Evans (2013, 438) using pottery data to argue the same case and saying this began in the Flavian period. James (2011, 155) adds that demand for 'Roman' goods would have been particularly strong among the military quarter masters and legionaries who saw themselves as Roman and yet lived (certainly by this date) for the most part distant from Rome. These goods helped distinguish them from the natives among whom they lived at the periphery of Empire. Similarly auxiliaries, raised elsewhere in the empire but based in Britain, would have encouraged trade with their own homelands to maintain their own identity. While the convergence of legionaries and auxiliaries in terms of their battlefield role in the later Empire may have changed this cultural vector, particularly with the emergence of *limitanei* border troops who became more embedded in the communities in which they lived, demand for military trade would have been maintained to support the *comitatenses* field army troops and *foederate* mercenaries.

Within this narrative of a largely prospering county until later in the occupation, Cunliffe (1969, 9) highlighted the differing fortunes of Rochester and Canterbury, particularly in the 2nd and early 3rd centuries AD. He said:

> *"Rochester remained surprisingly small and was only 23 acres in extent when, in the late 2nd century, it was enclosed by an earthen rampart and later a stone wall. The reason probably lay in its stifling closeness to London...Canterbury on the other hand developed into a densely packed town of more than 120 acres unhindered by the presence of a wall until the (late) 3rd century."*

One can see here, in the context of the Imperial economy (and even the provincial market economy), taxation and supply being aligned with a regional centre of government, in this case Rochester with nearby London. The respective populations of both towns are also instructive in considering their dimorphic development, with Goacher (2009, 391) again using the methodology of Millett (1990a, 185) to estimate that Canterbury's population at the height of the occupation was 9,500, while that of Rochester she estimates at 1,300. To provide context, such a figure would make Canterbury the sixth largest town in Roman Britain (London being the biggest at between 25,000 and 30,000 at its peak in the early 2nd century AD, Swain and Williams, 2008, 37), with Rochester remaining a classic small town with an administrative function on a key transport node. It therefore seems reasonable to me, especially given the preponderance of public buildings in Canterbury as discussed in Section 2.3.4, to agree with Cunliffe that while Canterbury was able to flourish as the regional *civitas* capital, Rochester's development was stunted given it fell in the shadow of the nearby provincial capital of London, a comparative short distance to the west along Watling Street and the Thames Estuary.

Fleming (2010, 2) is particularly impressed by the vibrancy of the economy of Britain during this early period (both in an Imperial and provincial context), commenting on the great distances which huge amounts of modestly-valued goods travelled from the core of Empire to the margins (an indicator of the levels of evident market integration here at this time). She explains that such levels of trade were higher during this period than at any time in the next 1,500 years. By the middle of the 3rd century AD however a major change is evident which occurred across all levels of society in Kent and indeed Britain. The debate regarding this change was initiated by Reece (1980, 77) who in his reappraisal of the 'Crisis of the 3rd Century' argued that around this time there was a gradual reversal of the flow of the Roman economy from a model where the centre supplied the regions to one where the regions were much more independent economically. Millett (1990a, 127) developed this theme, arguing that in the early period there would have been a major flow of wealth from taxes raised in the centre of Empire to the expanding edge where the military was situated and where the money raised was being spent. He explained that this inflated prices at the periphery of Empire and, once the Imperial expansion had finished, created a local market where it then made sense for goods to be produced and procured locally given the additional cost of those supplied through inter-provincial trade. Mattingly (2006, 500) also acknowledges this change, saying that manufactured goods from the Continent were a rarity in Britain by the 3rd and 4th centuries AD (indicating a decreased level of market integration here). He adds that in his view one of the key reasons for this import-substitution (the replacement of Continental imports by regional goods) was the need to lower the cost of the always exponentially large military presence in Britain. Such views act as a useful counterbalance to arguments that the later regionalised economy was the result of the disruption of inter-provincial trade caused by periodic civil war and major barbarian incursions in Gaul, Germany and northern Italy in the 3rd century AD crisis. Halsall (2013, 89), summarising a traditional view of this debate, explains his view that:

> *"The production of Roman goods that had taken place in the heart of the Empire had moved out to the provinces. Instead of a unified system exporting Roman goods from the core to the periphery in return for raw materials flowing from the periphery to the core, a more fragmented system now emerged. Production and consumption took place within provincial units."*

For balance it is worth noting here that Mattingly (2006, 510) has a slightly different view, even while agreeing that the change in the flow of imports did take place. In this regard, rather than seeing the economic experiences of occupied Britain as chronologically bi-polar, he looks at the issue on three different levels. These are the regional economy as part of the wider Imperial economy (see Section 2.2.2), as part of the provincial economy (see Section 2.2.3), and additionally as part of the extra-provincial economy. Crucially, he says that he believes the development of the latter two were always under the domineering influence of the first, even after the flow of imports was reversed (2006, 496).

Even noting this slightly different approach, as can be seen all of the informed commentators referenced above see a change in the direction of the flow of goods into and out of the province to a greater or lesser degree by the mid-3rd century AD. It is perhaps in this light that we should view one of the most visible manifestations of such an adjustment in occupied-Britain, namely the significant changes in the exploitation of natural resources by the extractive industries in Kent and the South East. This is evidenced by the virtual disappearance of large scale iron manufacturing in the Weald, and the replacement of industrial scale ragstone quarrying in the Medway Valley by much more localised activity (both discussed in depth in Chapters 3 and 4 below). Change is also evident in the fortunes of the elites residencies in the county as identified by Blanning (2014, 480) in Section 2.3.2, this arguably relating to the disappearance of the larger state-run enterprises (see discussion in Chapter 5). As Fleming (2010, 5) explains: *"...many villas in Kent...doubtless the country homes of families who had prospered in London's good times, were deserted."*

Figure 10: Decorated ragstone from Roman villa at East Farleigh, found by author re-used in a neighbouring garden wall. Simon Elliott.

As with elsewhere in the *diocese*, some elite dwellings were to reappear within the new town walls of urban centres when a move away from public buildings and industrial quarters towards orchards, market gardening and waste ground is visible (again though, less so in Canterbury). Halsall (2013, 92) says that by the beginning of the 4th century AD: *"...British towns were rather different from their 2nd century precursors. Where there had been shops, larger town houses appeared."*

Not all of the elite villa residences were abandoned in the middle of the 3rd century AD however, though even for those that remained (such as that at East Farleigh in the upper Medway Valley, see Figure 10) change was coming. Moorhead and Stuttard (2013, 205) say that by the beginning of the 4th century AD there is strong evidence of a further change in the ownership of those that survived, they citing examples in Kent. This event is particularly evident in the Medway Valley, for example at the villa sites at Eccles, The Mount in modern Maidstone, East Farleigh and Teston. In this regard, the coin hoard of 'Tetricus' found near Allington Castle could point to trauma in the county following the failure of the Gallic Empire (though, for balance, it could also be related to severe coinage devaluation at this time, Kulikowski, 2016, 180), while other events leading to such changes may have been the Carausian revolt and its subsequent failure in the late 3rd century AD, and the later usurpation attempt of Magnentius in the 4th century AD. In the aftermath of the latter Constantius II dealt with the *diocese* of Britannia particularly harshly. Such falls from grace for the elites were not uncommon, with Brown (2012, 188) highlighting the case of the late Roman poet Ausonius whose family had lost their place in the nobility after the Gallic Empire failed.

Interestingly, the narrative of the occupation in Kent concludes with a final burst of economic activity of some kind in the mid-to-late 4th century AD as evidenced by numismatic data, though at a time when Blanning has highlighted that only a quarter of the villa estates across Kent were still in use (2014, 480). Throughout the county, at surviving villa sites and at specific locations in the county such as Whitefriars in Canterbury (Moorhead, 2012, 2) and the water mills at Ickham (Brickstock and Casey, 2010, 75), the peak of coin activity is during this late period, between Reece's Period 17 (House of Constantine II, AD 330-348) and Period 19 (House of Valentinian, AD 364-378) for villas, Period 13 (Gallic Empire, AD 260-75) and Period 19 for Whitefriars, and Period 13 and Period 21 (House of Theodosius II, post AD 388 and the final Period) for Ickham. The juxtaposition of a peak in numismatic data with the continued decline in elite settlement is particularly interesting here, with the former arguably indicating societal stress as wealth was hoarded or discarded at a time of crisis. Another interpretation could be to reinforce the view however that the state was increasingly important economically in the region as the end of the occupation approached, with the roles of the elites diminishing and that of state officials and the military increasing, and with the peak of coin activity being explained by loss rather than hoarding or discard.

2.3.6 The End of the Occupation in Kent

The end of the Roman presence in Kent and the South East was ultimately wrapped up in that of the entire later *diocese*, with debate still raging as to what actually happened politically and economically in Britain as the occupation neared its end. Arguments range at one extreme from the *diocese* suffering a catastrophic collapse (sometimes with a transitional phase), to a belief that the change was gentle in nature with the life experiences of those living through it being different but no worse.

Very bullishly in the apocalyptic decline camp, Reece (1980, 78) saw the decay starting early and talks of the disappearance of British towns by AD 350. Faulkner (2000, 120), writing from a radical perspective viewing the Roman Imperial project as an example of violent larceny on a grand scale, sees this decline starting even earlier, saying that: *"In...the mid-3*rd *century, civil construction-work in the major towns of Roman Britain all but collapsed...even basic maintenance and repair of the existing stock was sometimes neglected."*

He adds (2000, 130) that this urban degeneration continued dramatically into the 4th century AD, aside from a few oases of civilization amid the urban decay, leading to a final cultural collapse between AD 375 and AD 425. He details (2000, 148) that the rural experience paralleled this, being one of earlier boom turning to later bust, and believes that talk of a late occupation-wide revival should be rejected with the 'golden age' towns being replaced by a gloomy, urban experience in an age of blood and iron (2000, 130). While accepting the dramatic decline model, Faulkner's dialectic opposite de la Bédoyère (1999, 14) fundamentally disagrees with the former's long-held grim view of late Roman Britain. He believes the experience up to the point of collapse was very different, as evidenced by the late economic boom visible in archaeological data from the South West (though not in Kent as detailed by Blanning, 2014, 480). Of Faulkner's dystopian view of late Roman Britain, de la Bédoyère says: *"...It was never like that. The Roman Britain of the 4th century would still have been identifiably Roman to any Roman from any time."*

He sees a late cultural revolution in occupied Britain prior to a sudden collapse, with the *diocese* being a safe haven from the tribulations on the Continent and where an estate-based rural lifestyle for societal-elites replaced the previous urban experience.

Interestingly, Faulkner also talks of a transitional period between the later Roman occupation and the early period of Anglo-Saxon expansion during the 5th century AD, when he sees evidence for a potential peasant revolt (2000, 178). Esmonde Cleary (1989, 204) also talked of this transitional period in his discussion on continuity and change, highlighting a post-Roman but pre-Anglo-Saxon phase in the 5th century and arguing for a degree of acquiescence on the part of this post-Roman society with regard to the eventual dominance of Anglo-Saxon material culture throughout much of Britain. Brown (2012, 393) also highlights this transitional phase, arguing that late period regionalism was more marked in Britain than elsewhere in the Western Empire, including the possible re-emergence of pre-Roman tribal identities. Though such ideas of a 'Celtic Revival' remain contentious, it is noteworthy that many early Saxon military burials are to be found along pre-Roman tribal boundaries at sites such as Dorchester-on-Thames in Oxfordshire, with Yorke (1995, 30) and Henson (2006, 64) arguing that these reflect the deployment of *foederates* by these newly re-emergent tribal polities.

Moving to the arguments in favour of more gentle change, with no abrupt break or intervening transition, some archaeologists and historians have recently begun to argue that this later period was actually (for better or worse) one of gradual shift, where for many inhabitants the experience was not of an overnight catastrophic collapse but instead a slightly different experience as each generation came and went. In this regard, Gardner (2007, 257) talks of change being achieved in many areas not by rebellion or disaffection but by local communities exercising agency, generation by generation, to affect small changes. Under this hypothesis, for the majority of such people these changes would have been barely perceptible as they carried out their everyday lives, with Halsall (2013, 254) saying: *"Effective Imperial power was retreating but no-one knew they were living through 'the end of the Roman Empire."*

Such views offer a different interpretation to the black and white of Faulkner and de la Bédoyère. Developing this picture of gradual rather than radical transformation at the end of the occupation, evidence of the changing nature of the local economy is also visible in the archaeological record. Fuelled by a proliferation of domestically produced, Roman-style goods, Fleming argues (2010, 21) that actually the period from AD 290AD to AD 360 was a high-point of the occupation in certain parts of Britain (for example the South West) rather than Faulkner's calamitous ruination of occupied Britain.

This process of gradual change (as opposed to dramatic collapse) could be viewed as the natural result of the late Romano-British elites wanting to retain the trappings of *Romanitas* while taking a more independent approach to engagement with the central authorities of the Empire. It that regard Romano-British society would have seen less and less return for the taxes they paid, for example in the form of security against what had become endemic raiding. They had therefore begun a process of disengagement well before the early 5th century AD when Roman authority is widely accepted as ending. As mentioned above, Gardner (2007, 257) succinctly describes this as a decision taken by numerous communities at local level, independently over time, rather than a grand occupation-wide process. It should be noted that the central authorities of the Empire had long noted the risk of revolt in Britain given its remoteness from Rome and the large military presence, hence the gradual division of the original province of Britannia into two, then four and finally (possibly) five provinces as detailed above in Section 1.1. While this reflected an Empire-wide process, especially following Diocletian's Tetrarchic reformation, Britain can be seen as an extreme case given its unique local issues.

Whether the end of the occupation was revolutionary or evolutionary in nature, its visibility in Kent and the south east of Britain is not in question. Particularly illuminating is the data from regional coin finds which fall off dramatically as the very end of the occupation approached, with Period 21 Theodosian coins increasingly few and far between and at specific locations only. This is a pattern reflected across Britain, with Moorhead (2012, 212) saying that evidence from the Portable Antiquities Scheme (PAS) database shows the monetary economy was shrinking. Bronze coins of Honorius (Augustus in the West from AD 393) seem to be the last base type to arrive in the diocese in any quantity, though Moorhead (2014, 206) adds here that even these diminish to a trickle (from Aquileia and Rome) following the closure of the Gallic mints at Arles, Lyons and Trier in AD 395. Meanwhile, the last major silver issues in the West were *siliquae* of Arcadius (Augustus in the Eastern Empire from AD 383) and Honorius, minted in Milan between AD 397 and AD 402. Further, very few gold *solidi* (from the mints at Ravenna, Rome, Milan and Aquileia) seem to arrive in Britain after AD 408, one exception being a *solidus* of Jovinus (AD 411 – AD 413) recorded in the PAS database (Moorhead, 2014, 211). This numismatic insight is illustrative of a dramatic late fall in activity across the South East, highlighting the comparative rarity of coins of Honorius compared to Theodosius 1 (Eastern Emperor AD 379 to AD 392, also of the West to AD 395) and Valentinian II (Western Emperor AD 375-AD 392) at villa sites such as East Farleigh (one Honorian coin) and Teston (one Honorian coin). The only exception is the huge bronze coin hoard found at Richborough (excavated between 1922 and 1925) where 998 of the coins date to the reign of Honorius, but even here none are later than AD 402. This has been interpreted as being the final, for some reason unsued pay chest to arrive for the garrison there at the very end of the occupation.

Further refining our knowledge of Kent and the south east of Britain at the end of the occupation through numismatics, Moorhead (2014, 201) shows that very few of the late coin issues have been found outside of late period 'islands of activity' in the east and north west economic regions (significant industrial activity in the Weald having finished in the middle of the 3rd century AD). The vast majority of these coins were of base metal, these being indicative of military/ state activity given their use to pay the military and officials (see Section 2.2.5), they being far more common at late urban/ military sites in the 'islands' than at rural sites. Further, the increase in the numbers of clipped silver *siliquae* in this late period could plausibly be interpreted as additional evidence of a state/ military presence, the clippings from these coins under such a hypothesis being used to make bullion ingots (Moorhead, 2012, 211) to pay the state or military officials in the *diocese* at the very end of the occupation, with Robertson (1974, 34) having said that this left the clipped coins in circulation only at the few places they were still of relevance. Looking at the end of activity in the two 'islands' themselves, the disappearance of late coins shows the end of activity at various specific sites before the overall end of the occupation, at locations such as the Saxon Shore fort at Lympne. Similarly, using the same data, both Dover and its Saxon Shore fort, together with settlement around Folkestone, also fell out of use before the end of the occupation. Mattingly (2006, 388) more broadly highlights the abandonment of coastal villas in the region at the same time.

The overall effect of this pattern of decline was to leave Canterbury and Richborough as the 'last men standing' in the eastern 'island' in the region, as the surviving population retrenched into their core areas (Blanning, 2014, 480), though the decline continued with Canterbury seeming to collapse at the beginning of the 5th century AD as evidenced for example by the hurried Stour Street family burial dug into a roadway in the temple precinct (Brookes and Harrington, 2010, 28). A significant period of time then ensued with the town being to all intents and purposes abandoned before its later occupation by arriving Saxons (Millett, 2007, 183). This would have left Richborough as a lone beacon of the Empire in this eastern 'island', evidenced for example by the coin hoard detailed above, this site relying for survival on its strategic location facilitating access to the fading *diocese*.

Meanwhile, in the north western 'island' in the region, the data presents a similar picture to that in the east, with the surviving villas gradually falling out of use to leave only a few of the major ones in existence. The latest activity in the upper Medway Valley, as evidenced by coins of Honorius and Arcadius being found, is interestingly in the upper Medway Valley at East Farleigh and Teston (see Figure 11). These coins are again of base metal, hinting once more at a state or military presence. When the end did finally come though, it seems these final sites were abandoned in haste given that there is little evidence of the robbing out of useable building materials (though noting the late antique shift towards timber usage in building construction) as seen in the downriver Medway Valley villa sites, for example at Eccles, and in the Darent Valley. Further evidence of a rapid abandonment of the Teston site is provided by the fact that it seems to have fallen from local knowledge rapidly, given it has no association at all with modern field boundaries and, today, a mature tree line passes through the length of its main range. Meanwhile, the presence of early Saxon pottery found at Teston amid the final occupation phase can reasonably be interpreted

Figure 11: The end of the Roman villa at Teston, domestic range. Neatly stacked imbrex roof tile was found here ready for transport elsewhere before the site was rapidly abandoned. Simon Elliott.

to indicate the cause of this haste (this and other potential explanations are discussed in full in Section 4.1.4).

Taking all of the above evidence into account, all indicators point to a continuing trend of marked change in occupation period Kent in the late 4th century AD, except in the 'islands', and even here *Romanitas* fades as we enter the 5th century AD. By this time memories of an industrialised county with its intense industrial enterprises had long faded. What followed has been described as dramatic and grim by Moorhead and Stuttard (2012, 249), with Moorhead (2014, 11) later arguing that coin use ceased in Britain around AD 430.

By this point post-Roman British society was experiencing a brutal, sharp flattening out to coincide with the final disappearance of *Romanitas* (with the exception perhaps of the South West for a generation or two).

2.4 Industrial Activity during the Roman Occupation

While acknowledging the importance of agriculture in the pre-modern economy, industry nevertheless still played a highly significant and integrated role in the Roman economy and is indeed central to the core theme of this research regarding the exploitation of natural resources by the extractive industries in occupied Kent and the South East. Evidence of these processes can be perceived by us today in a variety of ways, for example the widely recognised data showing a high concentration of pollutants from Roman industrial activity (particularly lead and copper emissions) found in Greenland ice cores. Here, the only other major pre-later 18th century peak occurs during the 11th century and relates to industrial activities by the Sung Chinese on the other side of the globe (Borsos *et al*, 2003, 5).

Of course, industry did exist in these islands before the arrival of the Romans. Examples include the minting of coins in the LIA (for a Kentish example see the reference to a mint in pre-Roman Rochester in Section 4.1.4), pottery production, quern stone production, mining and metal production. Technological innovation also occurred, for example with the adoption of the potter's wheel, the rotary quern and the lathe. However, with the arrival of the Romans something truly revolutionary happened, certainly in terms of scale, engineering innovation, the presence of manufacturing and the growth of consumerism in response to the availability of newly mass produced goods (Gardner, 2013, 7). These new industries did not exist in isolation either but were features in a complex international economic system, and were supported by an equally complex maritime and road-based transport infrastructure. This allows us to consider such activity as being part of a much wider industrious landscape, reviewed as a whole across the islands of Britain here to provide context for the Kentish experience detailed in the two regional analyses below (S. Elliott, 2014b, 44).

The Roman economy featured industries both great and small. These ranged from epic state-controlled mining and quarrying enterprises (such as the *metalla* considered here) and manufactories producing a wide variety of products (for example weapons of uniform quality and size and *garum* fish sauce), through to local milling and food production enterprises. With the arrival of the Romans in Britain, this industrial suite then became a feature of the British experience of the Empire.

In a specifically British context, we can look at a variety of examples of this new industrial experience. These included industries which are specifically detailed in my own PhD research, for example large-scale iron producing enterprises (see Figures 12 and 13), tile and brick production, mill and quern stone manufacturing and the associated milling industry, and industrial scale mining and quarrying.

Another industry which thrived during the occupation was pottery production, and here Peacock's key 1982 work *'Pottery in the Roman World'* most usefully provides a model not only to view this industry in occupied Britain and elsewhere in the Empire, but also through analogy other Roman industrial activity. In his model Peacock broke down this crucial activity into a hierarchy of seven different modes (1982, 8, these covering both domestic and imported ceramics), starting with household production at the lowest end (the least visible in the available data), followed by household industry (evidenced by Dorset

Figure 12: Roman cinder-heap, Gallants Lane iron-working site, upper Medway Valley. Simon Elliott.

Black-burnished ware, known as BB1), individual workshops (Severn Valley ware), nucleated workshops (with urban examples including those in Colchester and rural ones the Alice Holt potteries), then manufactories (imported Samian ware), estate production (where amphora manufacture for estate goods would be an example) and finally military or official workshop production (Holt in Cheshire). For manufactories he did draw one link with the present which would sit comfortably with a modern economist, namely the evident connection at such sites between capital and labour, while at the other end of this spectrum between past and present commonality he identified the latter two modes of production (estate production and military/ official production) as ones which do not sit comfortably in his wider hierarchy, let alone in a modern context. Interestingly, he also precluded pottery factories in a Roman context (an eighth mode type, above the manufactory level) as a mode of production given the Empire's arguable failure to exploit mechanical power to any extent (1982, 10).

Focusing on the British experience of the pottery industry, here were found a wide variety of styles from differing modes of production (see Peacock's examples above, 1982, 8), catering for requirements high and low, these being a key indication of the arrival and ensuing spread of *Romanitas* in the islands. In this regard de la Bédoyère (2000, 9) states that Roman style pottery found its way into all areas of daily life, as can be seen in works of synthesis on Roman pottery types such as Tyer's atlas or the work of Willis on samian ware (Tyers 1996; Willis 2005; 2011).

To this native Romano-British pottery we can add a modest indigenous glass production industry, with major known glass production facilities including that located on the south side of the *forum* in London and at Caistor-by Norwich (Roman *Venta Icenorum*, Jones and Mattingly, 1990, 216).

Figure 13: Roman lead ingot, late 1st century AD, Gallants Lane iron-working site, upper Medway Valley. Simon Elliott.

Most recently Howell *et al* (2013, 10) have also highlighted further significant glass production in Roman London at the Bow Bells House site south of Cheapside, this being one of many such sites coming to light in the provincial capital. Meanwhile, occupied Britain also featured a regionally focused mosaic manufacturing industry which seems to have particularly thrived in the 4th century. Jones and Mattingly (1990, 224) argue in this regard that there were six specific mosaic schools in this later period.

Britain was also home to a thriving textile industry. In particular, the province was known within the Empire for two specific textile products. These were a type of the *birrus* rain-proofed hooded cloak, and a form of fine quality *tapetia* woollen rug (Wild, 2002, 1). In the AD 301 Edict of Diocletian the British version of the latter is actually listed as the best available across the Empire. For its cloth fibres this industry made use of sheep's wool and flax, with the occasional addition of hemp and animal hair. Silk is also present, though as an import. Some of the cloth would also have been dyed, with around 20% of the woollen products found at the northern border fort of Vindolanda showing evidence of dyestuff (Wild, 2002, 1). The dyes identified here were either imports such as Madder (*rubia tinctorum L*), which gave a red colour, or local lichens which gave a purple colour.

Brewing was also a major industry in occupied Britain, with Kent providing an excellent example of the industrial scale of brewing operations during the Roman presence in the islands. Carruthers (2014, 143) explains: *"In the Ebbsfleet Valley at Northfleet villa, brewing appears to have been taking place on an... industrial scale... Malting ovens, a barn and three brewing tanks with the largest holding up to 16,000 pints (9,092 litres) were excavated."*

Meanwhile salt (vital as a preservative and a flavouring for food) was also the subject of industrial activity in occupied Britain. Roman salterns of significant size have been located around The Wash (a continuation of a strong LIA tradition), the East Anglian Coast and Thames estuary, along the south coast, and around Bridgewater in Somerset (see discussion in Section 2.2.4 regarding the provenance of salt manufacturing sites as Imperial Estates). Further, brine springs associated with salt production have been located at Northwich, Middlewich, Henhull and Whitchurch along or near the Rivers Weaver and Dane, and around Droitwich in Worcestershire (Woodiwiss, 1992, 183). Even a type of fish sauce such as *garum* may have been produced in occupied Britain, with Biddulph (2013, 20) suggesting Stanford Wharf in Essex as a possible location while Locker (2007, 151) similarly suggests a site excavated at Peninsula House in London as another (the latter based on the remains of vast quantities of young herrings and sprats found there in association with shards from amphora used to transport fish sauce).

Last but not least, an occupation-period coin minting industry is also evident in Britain (this a function of state production set within the Imperial economy, see Section 2.2.2 above). Contemporary copies of Roman coins were produced here in very large numbers in the Claudian and Neronian period (AD 41-64), Severan Period (AD 193-235), the Barbarous Radiate period (AD 275-85) and in the 4th century from AD 330 to 348 and AD 355 to AD 364. Meanwhile the principal official mint for purely indigenous coins was first founded in London in AD 286 by the usurper Carausius, with coins being produced here from AD 286 to AD 324, and then AD 383 to AD 388 (Moorhead, 2014, 32). Of the 29 major Roman mints from across the Empire in the PAS database of Roman coins found in Britain, the 2,987 coins made in London make this mint the fifth largest represented, as one can see from Table 2.3 (impressive when those with higher representations are Rome, Trier, Arles and Lyon). Additionally, Carausius set up a second mint (dubbed the C mint) which manufactured further coinage, of which 417 are recorded in the PAS database. There were clearly far more coins being minted and in circulation in Britain during the occupation than both before or indeed after.

Table 2.3: Mints across the Roman Empire

Mint	Coins
Alexandria	47
Amiens	193
Antioch	108
Aquileia	474
Arles	4,300
"C mint" (Carausius and Allectus)	417
Carthage	11
Constantinople	50
Cyzicus	65
Gallic mints (Gallic empire)	1,949
Heraclea	48
London	2,987
Lyon	4,012
Milan	10
Nicodemia	37
Ostia	15
Ravenna	15
Rome	11,701
Rouen	6
Serdica	6
Sirmium	19
Siscia	712
Spain	45
Tarraco	5
Thessalonica	110
Ticinium	143
Trier	10,616
Tripolis	5
Viminacium	1
TOTAL	**38,441**

Number of coins from major Roman mints on the PAS Database, after Moorhead, 2014, 32.

A final point to consider here is that industry also had an impact on settlement, for example in the location of small towns and even the location of villa estates. While the latter are often associated with agriculture, I discuss below in Section 4.3 and Section 5.1.4 whether many of those in the Medway Valley were associated in some way with the ragstone quarrying industry. Other recent examples include the villa located by Wardell Armstrong Archaeology at Emerson's Green near Bristol. This has been associated with textile manufacturing and specifically textile dyeing (Symonds, 8, 2014).

2.5 Maritime Transport in the South East during the Roman Occupation

As referenced above in Section 2.2.1 by Lavan (2014, 1) and Bonifay (2014, 557), maritime transport (by sea,

Table 2.4: Comparison of transport options

Vessel/ Animal/ Vehicle/	Fuel Type	Distance able to carry 1 ton on 1 gallon
Roman Merchant Ship	Food/ cooking fuel	1,280 miles
Roman *codicaria* (towed river barge)	Food/ cooking fuel	32 miles
Mule	Fodder	2.4 miles
Roman ox-wagon	Fodder	0.8 miles

Selkirk, 1995, 144.

canal and river) was essential to the smooth running of the Roman economy and had a direct impact on the prosperity of a given region. In this regard it was therefore a major factor in the success or otherwise of industry in Britain during the occupation, including those of an extractive nature exploiting natural resources in Kent and the South East under review in this research. It is therefore discussed in detail here given it is a thread which runs through both regional surveys which follow.

Maritime transport was the preferred choice of transporting heavy goods over long distances in the pre-modern world, and even today, specifically because of cost given it is significantly cheaper than land transport. The wider importance of transport costs is illustrated by Russell (2013a, 95) who cites the example of the Baths of Caracalla in Rome where 50% of the construction costs were taken up by shipping and haulage. Therefore any advantage regarding this cost would be taken, and in this context Table 2.4 by Selkirk (1995, 144) is particularly instructive.

Hard occupation-period data for this transport price differential is embedded in the Edict of Diocletian which highlighted sea travel as the cheapest means of transporting goods, then inland waterways, and finally (by a distance) roads. With regard to the transport of Egyptian papyrus, figures in the Edict show waterways being 4.9 times more expensive than travel by sea, with roadways being up to 56 times more expensive than by sea (Campbell, 2011, 216). More recent examples of similar economics can be found in 16th century accounts from Corpus Christi College in Cambridge which details it cost the same amount to move a load of stone 130km by water as it cost to move it 16km by land (Russell, 2013a, 96). Emphasising the point, Russell provides one further example, saying: *"As recently as 1962, limestone from Portland in Dorset was cheaper to purchase at Dublin (c.625km distant), to where it was be transported by sea, than at inland Birmingham (c.210km distant)."*

While the riverine transport of materials from the extractive industries exploiting natural resources in Roman Kent and the South East is central to the regional analyses below, the above examples make it clear that it was actually transport by sea which was by far the most cost efficient means of carrying heavy loads from one place to another. A specific example is highlighted by Lyne (1994, 540) where he shows that the Dorset-based BB1 and Thames Estuary-based Black Burnished Ware 2 (BB2) industries were one of the main suppliers of pottery to the military garrisons in the north of Britain into the 3rd century AD.

With regard to this maritime trade around the islands of Britain two interpretative models are particularly useful to provide context for our review of occupation period Kent when considering sea traffic around the British Isles. Firstly, Morris (2010, 1) argues that from the LIA through to the end of the occupation three specific regional maritime exchange systems existed. These facilitated the transfer of peoples and materials (which he terms 'connectivity', a manifestation of regional market integration) and their interaction with each other, though at any given time this was dependent on political and economic conditions. The three systems were:

- The Atlantic System, ranging from the Atlantic coasts of Britain and Western Europe to the Western Channel.
- The southern North Sea and eastern Channel System, ranging from the eastern Channel to the east coast of Britain (and centred around the Straights of Dover).
- The eastern North Sea System, linking settlement from the mouth of the Rhine to Scandinavia.

More recently Evans (2013, 433) has argued in favour of a two trade route model around the British Isles, one on the west coast and one the east. He argues that both were driven by the need to supply the military presence in the north, with the west coast route having as its principal channel crossing route that between Brittany/ Normandy and Poole Harbour. Evans (2014, 435) uses ceramic data to show that it was to the latter that continental imports of spices, olive oil, fish sauce, dried fruit and Samian ware arrived for trans-shipment along the south coast and then up the west coast. In this regard the principal occupation-period ports in the west would have included those at the legionary fortresses at Caerleon (Guest and Young, 2009, 97, see Figure 14) and Chester (Mason, 2001, 43), and the harbour at Bowness on the western extremity of Hadrian's Wall (Mason, 2003, 116). Meanwhile, Evans argues that the east coast route (more important to this research) featured London as its principal international emporium with its strong continental links to Gaul and the Rhineland. Archaeological data supporting this interpretation includes evidence for imports of spices, olive oil, fish sauce, Samian ware and other Gallic and Rhenish fine wares, and Noyon and Rhenish mortaria (with Dannell and Mees referencing London as a key emporium for the import of this elite pottery in their 'least cost' discussion, 2015, 80). Once again Evans says that such imported goods were then

Figure 14: Amphitheatre at the legionary fortress of Caerleon, south eastern Wales. Home of *legio II Augusta*, the site featured a significant military and civilian harbour located immediately above the amiphitheatre in picture. Simon Elliott.

trans-shipped north, this time along the east coast. Allen and Fulford (1999, 177) earlier demonstrated through their own analysis of regional ceramic exports that trade along the east coast was two-way in nature, with local Black-Burnished wares from the Thames Estuary joining the elite goods going north and then appearing along Hadrian's Wall and even on the Antonine Wall. By way of reciprocation, Dales Ware and coal of north eastern origin were then shipped back to the South East. In this regard the principal harbour on the north east coast was that at South Shields on the south bank of the Tyne, especially after it was greatly expanded to support the Scottish campaigns of Septimius Severus in the early 3rd century AD (Moorhead and Stuttard, 2012, 162, and Elliott, 2016, 157).

Looking specifically at Kent and the South East, and the shipment of goods deriving from the extractive industries therein, these would have fitted seamlessly into Morris' southern North Sea and Eastern Channel System and Evans' east coast trade route in their respective interpretive models. The vessels in use here followed a formalised schedule designed to make the best use of their time and carrying capacity. Such merchant ships were sailing a circuit between a variety of Continental and British ports, stopping off at Folkestone for example, where they would drop off some of their cargo of luxury goods and replace it with locally made Greensand querns. These would then be sold with other goods as the vessel stopped off along the Wantsum Channel, in the Thames Estuary (and onto the London emporium), and then around the East Anglian and North Sea coast.

Allen and Fulford (1999, 178) argue that while this trade along the east coast into the North Sea and onwards to the Continent was common throughout the occupation, it increased markedly from the Hadrianic period and then declined from the middle of the 3rd century AD. They continue that north Kent and south east Essex were specifically the origin of much of the material transported to the north (much of which Evans argues would have been trans-shipped through London, 2013, 433), and say that epigraphic evidence for the presence of units of the *Classis Britannica* during the building of Hadrian's Wall is evidence of the important role the regional fleet had in this east coast trade (see discussion in Chapter 6).

An interesting microcosm of the east coast trade route is presented by the recent discovery of four potential blank millstones or columnal bases (dubbed the 'Medway Stones' and fully detailed in Section 4.1.4 below) found recently on the bed of the Medway between East Farleigh and Tovil (S. Elliott, 2014c, 11). The stones are manufactured from greensand, most likely from the Folkestone region and thus in all probability shipped from East Wear Bay back to the Medway Valley on a vessel returning after dropping off a load of upper Medway Valley ragstone. Under such a hypothesis the ship was then wrecked before being able to drop off its greensand load and pick up more ragstone for shipment on to the London emporium or back once again to the east Kent coast. Data supporting such an argument is provided in Building 5 at the East Farleigh villa site where a quern stone of Folkestone Greensand has been found, though interestingly in a 4th century AD context (again see Section 4.1.4 below).

Meanwhile, Black (2013, 41) illustrates that trade from the Kentish coast was equally important travelling west, for example with regard to Wealden box flue tiles which he says would have been carried westwards by vessels operating through the east Kent ports (and thus fitting into Morris' Atlantic System and Evans' west coast route in their respective models). Such vessels would have arrived in the Weald with a cargo to sell of goods from London or north Kent, for example Canterbury grey ware pottery. Once unloaded this would have been replaced with Wealden pig iron and tiles before the vessels continued west. Such trade was clearly two-way in nature, with Lyne (1994, 133) highlighting the large quantities of Dorset-produced BB1 pottery which found its way to the east coast of Kent after being a prime cargo on the return journey.

Given the prominence of the *Classis Britannica* until its demise in the middle of the 3rd century AD (see discussion in Chapter 6), it is likely that the frequency of the regional maritime trade in both models would have become more problematic later in the occupation (noting Evans' comment on the decline of east coast trade from the middle of the 3rd century AD referenced above). What is clear however is that maritime trade did continue, and with the possible involvement in some form of the state (again see discussion in Chapter 5). In this regard Moody (2008, 154) points to the appearance in the later 3rd century AD of imported BB1 in Thanet of a type most often associated with the military. We can also look here again to Building 5 at the East Farleigh villa, with its 4th century AD Folkestone greensand quern, and also another of millstone grit from Derbyshire or Yorkshire found in the same chronological context, illustrating the continuance of long-range trading.

A final point to note here is the cosmopolitan nature of those engaged in the maritime trade networks around Britain at all stages of the occupation. A recent find of an individual involved in these types of activities is that of a dedication stone (of Turkish marble) seemingly dating to the Antonine period naming a mortix *Tiberinius Celerianus*, a maritime merchant who we know of from a temple dedication to the Spirits of the Emperors found during excavations in Tabard Square, Southwark. In this epigraphy he emphasises at the same time his trade, his loyalty to the Emperors, his allegiance to his tribe and community, and finally to his home city (Killock, 2009, 47). Indeed, his is the first reference we have in the historical record of someone calling himself or herself a Londoner.

2.6 The Military Presence in Britain during the Roman Occupation

Given the importance to this research of the military presence in Britain during the occupation, and particularly the *Classis Britannica* regional fleet, I provide here an overview to inform the two later detailed regional surveys and following discussion.

2.6.1 The Size of the Military Presence

In terms of land forces, the initial Claudian invasion force under Aulus Plautius totalled up to 40,000 men (Grainge, 2005, 111) comprising four legions (legio II *Augusta*, legio IX *Hispana*, legio XIV *Gemina* and legio XX *Valeria Victrix*) and 20,000 auxiliaries carried by 900 ships (Grainge, 2005, 129). It is worth noting the huge size of this force (which does not count the thousands of servants and slaves who would have accompanied it), reflecting Caesar's experiences in his landings in 55 BC and 54 BC. With regard to the first of these, Caesar arrived with just two legions (legio VII *Claudia Pia Fidelis* and legio X *Fretensis*) totalling some 10,000 men, while the latter featured five legions (legio VII and four others) totalling 25,000 men plus 2,000 auxiliary cavalry carried in 680 military and 200 commandeered civilian ships (Mason, 2003, 78). Caesar's failures (if viewed as such) clearly had an impact on the Roman psyche, with Horace (*Odes*, III.v) referencing the unconquered and fierce Britons (along with the Parthians) as unfinished business in the early Imperial project of Augustus, he saying that: *"Augustus will be deemed a God, on Earth when the Britons and the deadly Parthians have been added to our Empire."*

In the case of both of Caesar's incursions his force was clearly insufficient, a matter definitively addressed for the AD 43 invasion (which itself can be argued nearly failed in the context of the Boudican revolt of AD 60/ AD 61).

With regard to the size of the military presence in Britain in the post-conquest period, Millett (1990a, 181) has calculated that the army totalled up to 20,000 troops. Mattingly (2006, 131) goes further, saying that it numbered up to 55,000 in the 2nd century AD, and between 20,000 and 25,000 in the 4th century AD. A useful statistic to consider here is that the British Army in early 2014 numbered just under 100,000 regular and reserve trained personnel. Therefore Mattingly's figure of up to 55,000 troops in the 2nd century AD is clearly a huge number for this northwestern archipelago of Empire, especially given a maximum population at the time of only 3,665 million (see Section 2.2.3 above) compared to the 64 million today. To provide even more context, Mattingly's military figure of 55,000 represents up to 12% of the entire Roman army at that time, in just 4% of the overall Imperial territory. It is also worth noting here that both Millett and Mattingly, in the same calculations which determined the size of the military in Britain, also look at the size of the associated garrison settlements, with the former estimating a figure of between 50,000 and 200,000 and the latter 100,000 in the 2nd century AD and 50,000 in the 4th century AD.

Now turning to the maritime component of the military presence in Britain, the *Classis Britannica* was the regional fleet which had as its sphere of operations the British Isles, the North Sea, the English Channel and the northern Continental coast as far as the Rhine (Milne, 2000, 127). It is considered in particular detail here in the context of the regional reviews in Chapters 3 and 4, and the discussion in Chapter 6 concerning its role in facilitating the regional extractive industries.

The *Classis Britannica* was one of eight regional fleets which came into being to join the two state fleets operating out of Misenum and Ravenna (*Classis Misenensis* and *Classis Ravennate*) in Italy. Its importance can be judged by the annual stipend paid to its commander when compared to other fleets, detailed in Table 2.5.

Table 2.5: Roman regional fleets

Fleet	Annual Stipend
Classis Ravennate	300,000 sesterces
Classis Misenensis	200,000 sesterces
*Classis Britannica**	100,000 sesterces
*Classis Germanica**	100,000 sesterces
*Classis Pannonica**	60,000 sesterces
*Classis Moesica**	60,000 sesterces
Classis Pontica	60,000 sesterces
Classis Syriaca	60,000 sesterces
Classis Nova Libica	60,000 sesterces
Classis Alexandrina	60,000 sesterces

Ellis Jones, 2012, 61. *Indicates a significant riverine component.

The fleet had its origins in the Claudian invasion of AD 43 when the naval force of 900 ships mentioned above was created to facilitate the incursion. Initially formed around a core of experienced men from the *Classis Misenensis,* the British regional fleet was not actually called the *Classis Britannica* until the Flavian period, when regional stability gave the province the opportunity to formalise its activities. By that time Cunliffe (2013, 386) speculates that the majority of its sailors and shipbuilders would have originated from the Morini and Menapii coastal tribes of Belgic Gaul and from the Batavi in the Rhine Delta, on the basis that these were the tribes in the region with the most experience in shipbuilding and maritime activity. Mason (2003, 31) earlier detailed that other regional communities who may have provided ship's crew included the Chauci, Frisi and Veneti, all well known for their nautical skills.

Boulogne functioned as the fleet's headquarters for much of the period of its activity, its base there being the *Classis Britannica* fort which enclosed an area of 12.5ha, just over half the size of a full legionary fortress (Mason, 2003, 30). On the British side of the Channel the principal centres would have included Richborough, Dover and possibly Lympne on the Kentish coast (Philp, 1982, 176), and Pevensey in East Sussex (Mason, 2003, 30). This significant presence on both sides of the Channel reflected wider Roman attitudes to the control of crossings, derived from the Empire being both expansionist by its very nature (at least until the early 3rd century AD) and also the economic super-power of its day, dependent as it was on extensive supply routes. This was a theme which Philp (1981, 100) readily developed based on the data derived from his extensive excavations between 1970 and 1977 in Dover. Of the Boulogne/ Dover crossing he said: *"...the importance of this route has been largely underestimated... It clearly formed a vital and indispensable extension...of the major communications network spreading up from the Mediterranean through Gaul to Britain."*

The manpower complement of the *Classis Britannica* can be inferred from the size of the wider Roman fleet, and its regional components, over time. The original Roman war fleets created to fight the First and Second Punic Wars were extensive, given the nature of the conflict across the western Mediterranean, numbering up to 60,000 men in terms of crew for the second iteration. Most of these would have been rowers, for example 30,000 such specialists (professionals and citizen levy, not slaves, Goldsworthy, 167, 2014) being needed for the 203 BC invasion of Africa to man 160 warships according to Pitassi (2012, 61). He believes that the 60,000 overall figure would have fallen to around 30,000 by the reign of Augustus at the turn of the 1st century BC, then rising to around 50,000 again by the reign of Hadrian in the 2nd century AD as the regional fleets reached maturity. Key factors in this increase over 140 or so years would have been the growth of the *Classis Germanica* (incorporated very late in the 1st century BC) and creation of the *Classis Britannica* in the later 1st century AD. As can be seen above, based on the stipend paid to their commanders, these two were jointly the third most important fleets in the Empire after the Italy-based *Classis Misenensis* and *Classis Ravennate.* It is from the compliments of these two latter fleets that we can start to infer the size of the *Classis Britannica*, with Mason (2003, 31) arguing that during the reigns of Otho and Vitellius in AD 69 the former had a compliment of 6,000 and the latter 10,000 (using a calculation based on the number converted for legionary land-based warfare during the civil war of that year). He then looks at the military mission of the *Classis Britannica*, arguing that it would need three squadrons to maintain sea control and communications, on the east and west coasts of Britain and in the Bristol Channel. Finally, he looks at the known fleet bases at Boulogne where he says the fort would have accommodated 3,500, Dover (640 men) and other bases such as Richborough, Lympne and Pevensey (the latter two both presumed to have featured *Classis Britannica* forts, though not yet confirmed). Based on all of these factors he calculates that the compliment of the *Classis Britannica* would have been around 7,000 men. Its main warships would he been *liburnae* bireme galleys, as illustrated in Figures 15 and 16 and fully detailed in Appendix C.

Figure 15: Roman galley, graffiti on waste lead from Caistor-by-Norwich, 2nd century AD. James Beckerleg.

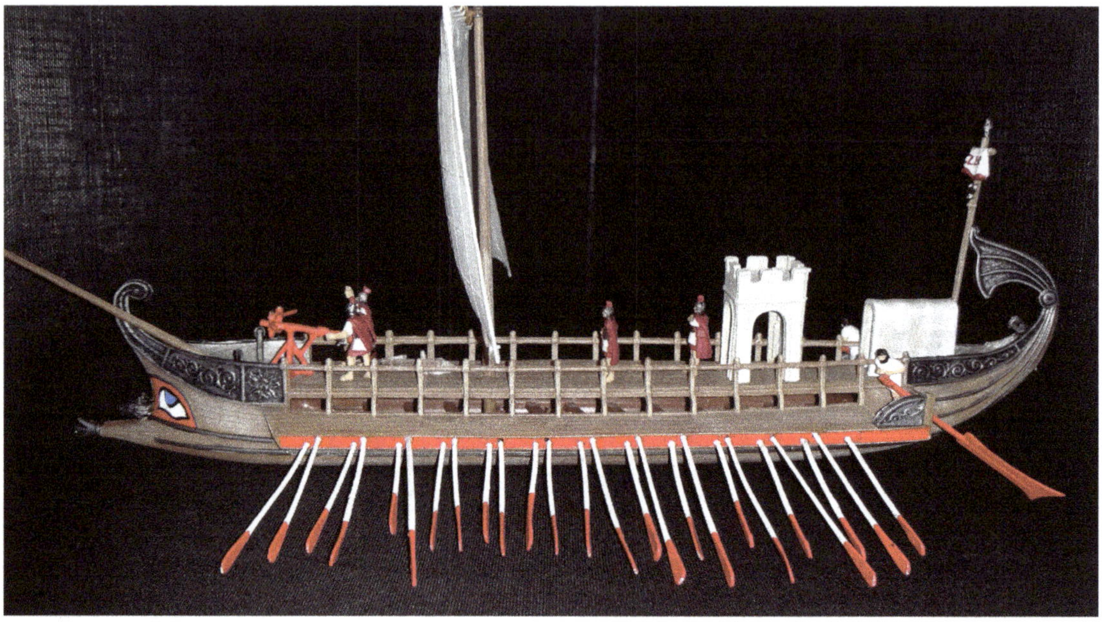

Figure 16: Accurate model made by the author of a Roman bireme *liburnae* of the *Classis Britannica*, showing standard layout including ram, ballista, mainsail and two banks of oars. Simon Elliott.

The last epigraphic testament to the existence of the *Classis Britannica* is that of *Saturninus*, ex-Captain in the British Fleet, dated to AD 244-249 (Russel, 2002, and see Appendix E below). Ingleton (2012, 10), Philp (1981, 115) and earlier Cleere (1977, 19) added that around this time the *Classis Britannica* fort at Dover also fell out of use, the latter saying it was deliberately and comprehensively slighted. Meanwhile Morris (2010, 147) also argues that around this time there is a marked decline in North Sea trade. Such data and anecdote has traditionally been used to argue the case that the *Classis Britannica* ceased to exist in the mid-3rd century AD, for example by Cunliffe (2013, 389). There is no clear data to indicate why the regional navy disappears at this time, though it is around this point that the 'Crisis of the 3rd Century' begins to gather pace (discussed in detail in Chapter 5). The fleet did survive the initial reforms of the military, and indeed separation of the original province into two, by Septimius Severus (Emperor AD 193 – 211) and his son Caracalla (AD 198 – 217), and the accession of Maximinus Thrax (AD 235 – 238). It may then have been the victim of the establishment fightback as a string of Senatorial-level Imperial candidates followed the assassinated Thrax (P. Elliott, 2014, 44). More likely though, its fortunes may have been tied to the fate of the Gallic Empire initiated by the usurper Postumus in AD 260 which featured Britain amongst its constituent parts (P. Elliott, 2014, 121). In this context, the regional navy could then have found itself on the wrong side of the power struggles of the Gallic Empire (of its four 'Emperors', three were assassinated) or indeed of the Aurelianic reconquest in AD 274. In the latter regard, it is instructive that the coin hoard of 'Tetricus' detailed above in Section 2.2.5 has been found at Allington Castle north of Maidstone on the site of a prospective Roman villa and quarry (see Section 4.1.4 below, particularly relevant given the discussion regarding the role of the *Classis Britannica* in the ragstone quarrying industry of the upper Medway Valley in Chapter 4).

Maritime operations after this time around Britain largely depended on resources specifically created to fulfil isolated regional requirements. Pitassi (2012, 15) says that by the reign of Diocletian (AD 285 – 305) the Imperial regional fleets had been replaced by smaller squadrons (reflecting changes to the army in the same period, with the *Classis Anderetianorum* being a potential British candidate here). Russel (2002) agrees that the military and state functions of maritime activity in Britain after this time become more regional, perhaps linked to individual military units. An example that a maritime capability did continue would be the 4th century AD *numerous barcariorum Tigrisiensum* detachment of Tigris boatmen (Finkle, 2014, 138), operating on the River Tyne from the fort at South Shields (Roman *Arbeia*). Additionally we can reference the later *limitanei/ riparienses* units which policed border regions featuring a river (for example the Rhine or Danube) and which would certainly have maintained a riverine capability (Hughes, 2010). Peaks and troughs in maritime capability seem to have become the norm in this later period, with Cunliffe (2013, 424) saying that at certain times, in certain circumstances, the North Sea would still have been alive with shipping. Moorhead and Stuttard (2012, 174, 207) believe that such highpoints would have been during the Carausian revolt, and later during Julian's period as Caesar when he ordered 600 ships to be collected and built to convoy grain from Britain to feed the Rhine armies. They go on to explain that this latter effort was only short lived, and that by AD 367 the fleet was once again too small to prevent barbarians using the seas around the *diocese*. Finally, Pitassi (2012, 21) argues that by AD 395 any semblance of control in the North Sea had been lost.

For our purposes though we can return to the 2nd century AD when the *Classis Britannica* was at its height, taking Mattingly's 55,000 figure for the Army and adding to it the

7,000 personnel of the regional fleet, giving a huge total of 62,000 military personnel operating in and around the islands of Britain. This would represent an eighth of the entire Imperial military complement at the time, a figure large enough to prompt Herodian (2.15) to comment on its size and power.

2.6.2 Mechanisms of Command and Control

Having determined the potential size of the military force in Britain, and how the *Classis Britannica* fitted into this, we can now explore the channels of command by which this would have been utilised, both for military activity and also as agents of the state for other functions (for example managing industry in occupied Kent, see Chapter 5 regarding the debate on this issue). After all, 62,000 personnel gives a very sizeable force of state employees looking for gainful employment, and though there were clearly spikes of intense military engagement during the occupation, this was not the case all of the time.

In terms of governance, a major Imperial Roman province until the Diocletian reformation had two different chains of command to ensure its smooth running. The first was the staff of the Imperial Governor, and the second that of the Procurator (Birley, 2005, 3). The former was tasked with ruling the province and was the chief administrator of Roman law. A consular status Imperial legate (senior officer, today a senior General), he would be supported by an *iuridicus* (legal expert), three legionary legates and three senatorial rank military tribunes. Equestrian-rank officers would make up the remaining military hierarchy, with 15 legionary tribunes and up to 60 auxiliary commanders (Mattingly, 2011, 219). Clearly, the military force of the province sat within the chain of command of the Governor.

However, it was the Procurator who was tasked with making the province pay, reporting directly to the Emperor's *fiscus* (exchequer) and thus being the main representative of the Imperial economy in the province (though also having a key role to play in the provincial economy). The term Procurator derives from the Latin verb *procurare* ('to take care') and referred to the individuals, usually freedmen, hired by the wealthy to manage agricultural or financial estates (Fuhrmann, 2012, 195). The use in an Imperial context originated in the reforms of Augustus (Birley, 2005, 298) as he began to establish the apparatus of Empire, with the name being applied at this senior level to those managing the Imperial private and public estates. The individuals so employed proved particularly useful in a provincial role given the comparative lack of Senatorial interest in this regard (Birley, 2005, 298). By the beginning of the reign of Claudius the first full provincial Procurators had appeared, for example in Judea in AD 41, they being independent of the Governor and thus acting as a useful political counterweight. Very shortly thereafter they also began to be recruited from equestrian ranks. The Procurators of Britain were equestrian-rank from the beginning (Birley, 2005, 298), with the lesser *procuratores* employed below them including other equestrians and also freedmen, some of the latter from the *familia caesaris* (the Emperor's freedmen and slaves). Note should be taken at this point that while there might appear to be parallels with a modern civil service here, we are talking about a strictly limited number of individuals so the analogy is not as useful as it might appear. The *procuratores* would have been registrars, finance officers and the superintendents and specialists tasked with running major state sponsored industrial activity (for example the *procurator metallorum* tasked with running the *metalla* state mines and quarries, discussed in full in Section 2.6.3 and Chapter 6).

In Britain the Procurator was styled the *procurator Augusti*, with the first holder of the post being Publius Graecinius from AD 43 to AD 60 (he, at least initially, also being the Procurator in Gaul). A key point here is that the Procurator in Britain was appointed immediately after the Claudian invasion, the Emperor intending to make the new province pay from the word go. After all, it had to be shown that it was worth the conquest as set out in Section 2.2.2, and this priority remained writ large in Britain throughout the occupation given the large military presence which always placed a burden on the regional economy.

The Procurator had specific responsibility for collecting taxes through the mechanisms set out in Section 2.2.2 which, as Pearson (2006, 39) explains, would have included responsibility for mines, quarries, Imperial estates and other state monopolies and where the superintendents and specialists detailed above would have been stationed. The territories attached to such industries would have held a special position within the province (Hirt, 2010, 106), very relevant to this research.

In terms of total numbers, taking both the Governor and Procurator's staffs together, this would give a combined number of less than 80 senior officials to run the province legally and financially. As Mattingly (2011, 219) says: *"Even if we include legionary centurions and unknown numbers of Imperial freedmen and Imperial slaves, the core bureaucratic team in the province was very small."*

After the Severan reforms in the early 3rd century AD with the creation of the two provinces in the islands of Britain (of *Britannia Superior*, with London as its capital, and *Britannia Inferior*, with York (Roman *eboracom*) as its capital), the division of command of the Roman military in the region is unclear, though common sense dictates that the troops on the northern and western frontiers would have reported to the Governor in York while those having a wider regional responsibility (for example the *Classis Britannica*) would have reported to the Governor in London.

Later, following the Diocletian reformation, the Procurator's role was replaced by a *vicarius* who had responsibility for the wider *diocese* (the original province being further subdivided), with each of the new smaller provinces in Britain being governed by a *praeses*-rank Governor (who combined the role of the original Governor and Procurator). With this reformation the constellation of civil servants supporting state activity would have been greatly expanded, given the additional tiers of governance.

The civilian role of the *Classis Britannica,* far more so than the legions, is central to this research and is discussed in detail in Chapter 5, so here I will dwell briefly on the instruments by which the fleet might have actually been utilized in this role. The Procurator was the most relevant official facilitating the use of military assets for civilian purposes (as part of the Imperial economy). Clearly there would have been some governing mechanism to allow him to call upon the assets of the Governor – the military – when they were not required elsewhere and when serious state-level resources were needed to facilitate his task of making the province pay its way. Hirt (2010, 199) believes this was through the appointment of a *beneficiarii procuratoris* by the Governor from the ranks of the military to serve on the Procurator's staff, acting as a go between. Meanwhile the importance of the regional fleet in Britain is directly indicated by the fact that the second most senior procuratorial position on the Procurator's staff was the *praefectus Classis Britannicae* (fleet Admiral, Birley, 2005, 298), emphasising the importance of access to the fleet's resources by the Procurator. By way of explanation here, each of the regional fleets' *praefectus classis* were the direct appointment of the Emperor, the post being part of the equestrian career structure. A very senior position, it was also rather bi-polar in that while the overall reporting line was through Procuratorial channels to the Emperor directly, when in theatre the *praefectus classis* and his fleet were under the local military command of the Governor given its military function. This link in the role of the *praefectus classis* between the civilian and military chains of command is even more explicit when one considers that we know of at least one individual, Marcus Maenius Agrippa L. Tusudius in the reign of Antoninus Pius (Emperor AD 138 – AD 161), who combined the post of Procurator with being the actual military commander of the *Classis Britannica* (Mason, 2003, 32). This intimacy reflects the fact that to facilitate the exploitation of provincial raw materials, the Procurator and his staff would have used whatever resource was to hand, and it is in this context that we should therefore not simply think of the fleet in a maritime context, but also as a tool to perform other functions. Parfitt (2013, 45) is explicit about this, saying: *"The Classis Britannica seems to have functioned mainly as some kind of army service corps, supporting the Government and provincial army, rather than as a Navy in the modern sense."*

2.6.3 The Military in a Civilian Context

Developing this theme, it is clear that the Roman military more broadly was used not just for conflict-related activity but also as a resource to be deployed by the state as and when required for other functions. This was after all an era before the advent of a civil service, nationalized industries or a free market capable of being engaged by the state to complete large-scale capital expenditure projects. This debate is important when we consider and interpret the data in the regional analyses below, especially with regard to the role of the *Classis Britannica* as discussed in Chapter 5.

In the first instance, given the chain of command outlined above, clearly the military was capable of providing an administrative capability in a given province, helping with the smooth running of the Imperial economy and ensuring the presence of the state within its provincial equivilant. A specific example is provided by one Babatha, an early 2nd century AD resident in the province of Arabia, whose private papers were found in Israeli caves at Khirbet Qumran. These detail her property assets as recorded in an official census in December AD 127, and indicate that her declaration was made to a cavalry commander named *Priscus* who was acting in this administrative role of behalf of the Governor (Goldsworthy, 2003, 146).

Another public function for the military was that of firefighter, with D'Amato (2009, 14) detailing as an example the two cohorts of military personnel permanently stationed at both Ostia and Puteoli (modern Pozzuoli) to guard against the occurrence of fire at the port facilities there.

Additionally the military were also highly skilled in engineering. In particular Goldsworthy (2003, 146) details the large number of specialist craftsman and engineers attached to military units. Blagg (2002, 182) adds that the former included highly experienced architects and surveyors. Additionally, the professional soldiers themselves were trained engineers in their own right, able to fully participate in the construction of not only their own fortifications but civilian structures too. To this end Connolly (1981, 239) explained that each legionary in a Principate legion had to carry a saw, pickaxe, sickle, basket, chain and leather strap. The same was true of the sailors and marines of the regional fleets, with D'Amato (2009, 15) detailing the remains of one such individual found during the excavations at Herculaneum who carried not only his military equipment (for example sword and dagger on a military belt) but also a bag of carpenter's tools. He has been identified as originating in the *Classis Misenensis.* Meanwhile, one of the most obvious examples of the engineering prowess of such troops, both specialists and regulars, were the roads built throughout the Empire. The commonality of their structure reflects the use of the military in their construction, even when they were not being built for specific military use (Goldsworthy, 2003, 146).

Other specific examples of the military carrying out purely civilian construction and maintenance are also evident across the Empire. One such is the aqueduct outside the *colonia* of *Caesarea Maritima* on the coast of Judea where we know from inscriptions that a vexillation from the legio X *Fretensis* was used for maintenance work. Topically for this research given its focus on riverine use, another example is the three-mile long canal with bridges constructed near Antioch in AD 75 by vexillations from four legions together with 20 auxiliary cohorts.

From construction we can turn to another manifestation of the engineering skills of the Roman military being used in a civilian context, namely mining (other elements of the workforce in the *metalla*, for example slaves, are considered in the relevant sections at Section 3.5 and Section 4.4 below given the focus in this sub-section on purely military matters). In this regard de la Bédoyère (1992, 100) is unequivocal that the majority of such extractive industries exploiting natural resources were under state control. Using data from mining operations across the Empire, Hirt (2010, 106) details that the actual mechanism of such state control would have been public ownership under the authority of the Emperor (see Section 2.2.2 and Section 2.2.4 above for the discussion on the various options available within the Imperial economy to actually control and manage the larger scale *metalla* operations). Whatever the ultimate such arrangements were, given the military were readily to hand, it was invariably the troops who initially facilitated each major mining operation. This was clearly common, a fine example being the award of triumphal honours by Emperor Claudius to Curtius Rufus, Governor of Upper Germany, for allowing his troops to facilitate silver mining (Goldsworthy, 2003, 148). Mattingly (2006, 507) adds another example of the military being involved in mining, this time for lead. This metal was an important part of the Roman economy, for use in its own right and also as a source of silver for coin production (in this latter case the silver being extracted from argentiferous lead by the process of cupellation, Jones and Mattingly, 1990, 185). The exploitation of lead during the occupation is a useful tool for the archaeologist given that it was produced in ingots or 'pigs' which were usually stamped and dated, giving insight into their origins both geographically and chronologically. We know the legions were producing lead as early as AD 49 in Britain because a 'pig' originating in the Mendips has been found at St Valury-sur-Somme in France stamped with the mark of the legio II *Augusta* and dated AD 49. This shows that the lead was not just for local use, and three other examples found in Britain from this period indicate that the export route was through Southampton Water (Jones and Mattingly, 1990, 184). Salway (1981, 634) argued that the early exploitation of lead in Britain was so important that it was a key factor in the earlier prioritization of the South West for conquest in Vespasian's famous campaigns in the mid-late 40s AD when legate of the legio II. In this region, at sites such as Charterhouse-on-Mendip, lead production soon reached industrial proportions (the lead here having a particularly high silver content), to be quickly joined by other areas such as Wales and Northumberland. In fact the industry was so successful that by the AD 70s Britain had surpassed Spain as the leading province supplying the metal, to the extent that the state directly intervened. As Salway (1981, 635, he citing Pliny the Elder, *Nat. Hist.* 34.49.164) explained:

> *"By Flavian times the much greater ease with which the surface deposits of Britain could be worked than the mines of Spain... had proved a serious embarrassment to the Imperial Government and production was limited by law, presumably for political reasons such as the protection of interests in other provinces."*

As detailed in Section 2.2.2, once successfully initiated by the military, the lead mining and manufacturing claims were quickly let to either *metalla* contractors, companies of *socii* investors or ambitious entrepreneurs (Mattingly, 2006, 507), though with the state perhaps retaining some kind of controlling interest as part of the Imperial economy. We have a specific example of one of these entrepreneurs taking over a lead mining *metalla* concession, the freedman C. Nipius Ascanius. His private stamp has been found on a Mendips 'pig' dated AD 59, and he is later found acquiring lead deposits in the early AD 60s in the Clwyd region in Wales, here before this district was actually pacified (Salway, 1981, 634). This is a useful counterpoint to the different experience of state involvement in the upper Medway Valley ragstone quarrying *metalla* discussed later in Chapter 5 where a longer term military presence is debated as an alternative to the early letting of commercial *metalla* contracts (again with reference back to Section 2.2.2).

The limit on lead production seems to have been lifted later in the occupation, with Todd (1996, 47) using data based on pottery analysis to show that lead mining continued to thrive in the Mendips until at least the 3rd century AD. The lifting of the limit may actually have occurred during the reign of Hadrian (Salway, 1981, 635) when lead mining in Derbyshire began, the Emperor's name appearing on 'pigs' from this source which indicates the state was again initiating production, though the industry here may not have been as successful as that of the South West given the comparatively poorer silver content. The same was true of lead mining operations in Shropshire and Yorkshire. Lead manufacturing did continue into the later period though, to facilitate demand for pewter in addition to its more traditional uses, and once again official stamps on 'pigs' indicate state involvement.

Meanwhile, another strong example of the state being involved in mining is with regard to iron ore and the associated iron manufacturing industry. This is specifically detailed below in the Chapter 3 regional study covering the Weald, with the role of the *Classis Britannica* in the extensive iron manufacturing industry there discussed in Chapter 5.

Staying with the Imperial economy, from mining we can move on to state involvement in quarrying *metalla*, of particular importance to the Chapter 5 Medway Valley regional analysis below, with once again the role of the *Classis Britannica* therein being discussed in Chapter 6. Other known British examples of military-run quarrying, in this case specifically early, include the various types of freestone for pre-Flavian memorials around fortresses such as Colchester, Gloucester, Lincoln and Alchester. The manifestation of the state in these cases were the legions and auxiliaries themselves, with Hayward (2009, 112) in this context saying: *"The army would have had the necessary specialists, manpower, equipment and organization at this time to survey, quarry and supply two metre long blocks (of freestone for monuments)."*

One can of course add here as a further example the 11 known quarries used to provide worked material, usually local sandstones, for the construction of Hadrian's Wall (Breeze and Dobson, 2000, 31). Here, in a number of cases, inscriptions in the quarries themselves identify the military units actually carrying out the stone extraction.

2.6.4 Military Infrastructure

Next for discussion, *fabricae*-style workshops in Britain during the Roman occupation are important in a regional context (and again particularly relevant to the discussion in Chapter 5). Such workshops, large and small, were vital to ensure the readiness for action of the military in all of its roles, and are thus an indication of an extensive military presence in a given region. Analogously such workshops fit specifically with Peacock's 'military and official' mode of production for the Roman pottery industry (1982, 11), though in terms of size the smaller would have matched the individual workshop mode with the larger being more akin to the civilian manufactories (or even factories if a degree of mechanization can be proved, see Section 2.4).

In the first instance there is a strong and recognized association for smaller *fabricae* to be located within legionary fortresses. Examples include Caerleon, Inchtuthil and Exeter, with geophysical data at the former highlighting a courtyard building featuring extensive burnt deposits which has been interpreted as the legionary metalworking workshop (Guest and Young, 2009, 105). At Inchtuthil in Perth and Kinross a similar structure, this time built from timber, has been identified as a small *fabricae* by data including a smithing hearth and a slag pit (Petrikovits, 1975, 93). Another similar structure within the walls of the legionary fortress at Exeter has also been interpreted as a *fabricae* (Bidwell, 1980, 35). Such workshops have also been found at smaller fortress locations, for example at Corbridge to the south of Hadrian's Wall on Dere Street. Here, archaeological data including arrowheads, iron scales and iron slag found alongside hearths and tempering tanks has resulted in one of the buildings within the site of the Agricolan-period fort being interpreted as a small military manufactory (Magness, 2011, 351).

On a larger scale however, perhaps at Peacock's nucleated workshop level (1982, 11), a similar dataset has been found in a non-fortress context at the Ickham Water Mills site on the Little Stour River to the east of Canterbury. Well known as the location of an extensive occupation-period water milling facility, it was one of only five such sites known from the occupation in Britain (Alexander, 2011, though noting Shaffrey's recent assertion that powered mills were far more common in occupied Britain than the current data suggests, 2015, 73). With four specific mills at Ickham known to have operated from the 2nd through to the 4th century AD (using millstones and querns manufactured both locally and also imported from the north of Britain and Gaul, Ridler, 2010, 251), other evidence now suggests that the site at Ickham had other functions. Excavated in the 1970s, material culture finds included five official lead seals, four stamped with the heads of late Roman Emperors (one of Constantine II, AD 337-340, and three of Julian, AD 360-363) while the fifth has a stamp indicating it originated from Smyrna in Anatolia (Young, 1984, 33). The excavators themselves additionally detailed the finding of spearheads, ballista bolt heads, *lorica hamata* chain mail and late period helmet components (Mould, 2010, 144), while Young (1984, 35) also highlighted the finding of late Roman fittings for *cingulum* military belts. Most enigmatic though is a large iron hammerhead found at the site featuring mechanical deformation on one side. The presence of this find, together with the seals, military equipment and large quantities of iron, bronze and pewter waste indicates that the water courses were also being used to power water hammers to produce or maintain military equipment to support the nearby military presence in Canterbury and the regional Saxon Shore forts. In this regard Millett (2007, 182) says: *"The location of this roadside settlement close to Richborough and with easy maritime communications suggests continuity in the importance of east Kent for military supply into the later Roman period."*

R.J. Spain (1984, 32), in his detailed study of Roman watermills in Britain, argued that the site was indeed being used by the Roman state to support the military, while Young (1981, 36) also gave some idea of scale, saying:

> *"Ickham...was most probably in the 4th century an official works depot for local units of the Saxon Shore, supplying them with flour and metal work. The presence of lead seals may suggest also that it was used for storage. It should not be regarded as a full-scale fabricae, the Imperial arms factories recorded (on the Continent) in the Notitia Dignitatum. There were few of them and, according to the Notitia, non were located in Britain apart from a clothing factory. A much better parallel would be the legionary works compound (detailed above) at Corbridge."*

This seems a reasonable interpretation based on the available data, with Mattingly going even further in suggesting an Imperial Estate interpretation (2006, 386 and 455), though more work is clearly required to definitively identify the site as any form of state-controlled military enterprise.

3

Regional Analysis – The Weald

This chapter is the first of the two regional surveys which form the core research of the book, focusing in this instance on the occupation-period iron manufacturing *metalla* of the Weald in Kent, East Sussex and Surrey. This was an industrial-scale operation which supplied much of the iron required by the military in the north and west of the province (later provinces), and more broadly for other purposes across Britain (including London) and for export, through to the mid-3rd century AD. The research presented here is specifically relevant to the debate about the state (in the form of the military) presence regarding this and other regional industries, in the context of the Imperial and provincial economies (and indeed the subsidiary research question concerning Imperial Estates) as set out in Section 2.2.1 through Section 2.2.4, these themes then being addressed in detail in the discussion in Chapter 5. The chapter begins with my detailing the key primary evidence sites in this region (including a site list of the most important as defined below), followed by a discussion on occupation-period settlement in the Weald, a similar discussion on the important theme of transport infrastructure in the region, then a detailed analysis of the Roman iron industry in the Weald, a reflection on the workforce therein, and finally a discussion about the occupation-period tile and brick industry embedded within the iron industry here.

3.1 Key Data: Details of Primary Evidence Sites

Here I set out the key data from all of the primary evidence sites in this region to facilitate the discussions in each following section of the chapter, preceding this site-by-site analysis with a discussion on the nature of the evidence, a review of the origins of the exploitation of natural resources in the region and a comment on site selection and grading, all to provide understanding for the reader in the ensuing site list at Section 3.1.4.

For general background, the Weald is the name used to describe the region between the chalk escarpments of the North and South Downs and is a major geological feature impacting on land use in Kent, East and West Sussex and Surrey. Divided into two distinct geological areas, the High Weald on sandstones in the centre and the surrounding Low Weald on clays, it has always been associated with heavy soils which are difficult to farm (Cleere and Crossley, 1995, 8). For much of the historical record the region has been heavily wooded and it is indeed from this phenomenon that it derives its historical name, Weald being a derivative of the Germanic *wald* (meaning forest, Everitt, 1986, 25). During the Roman occupation it featured a flourishing iron manufacturing industry and that is the major focus of this chapter.

3.1.1 Nature of the Evidence

I am fortunate that as a starting point there is already a mature body of work in existence and available for use given historical and archaeological investigation of the occupation-period Weald is about to enter its third century. This work has been led by some of the leading experts in their fields, for example in the 19th century M.A. Lower and J. Rock, in the 20th century E. Straker, I. Margary, C. 'Fred' Tebbutt, G. Brodribb and H. Cleere, and most recently D. Rudling and J. Hodgkinson. Research continues to be active given the proliferation of regional historical and archaeological organisations, for example the Wealden Iron Research Group (WIRG), HAARG, the Sussex Archaeological Society (SAS), the Battle and District Historical Society (BDHS), the Independent Historical Research Group (IHRG) and the Kent Archaeological Rescue Unit (KARU). We are also fortunate that data from the regional investigations is regularly published in key regional historical and archaeological publications such as *Archaeologia Cantiana, Sussex Past and Present*, *Sussex Archaeological Collections,* and *Wealden Iron*, with now defunct publications such as *Sussex Notes and Queries* providing an additional treasury of historical site investigation reports from the 19th and 20th centuries.

All of the above have served to challenge a view among some archaeologists and historians that, given its comparative inaccessibility and lack of agricultural activity compared to other areas of the South East, the Weald is a region reticent to yield its archaeological secrets (Brandon, 2003, 36). Indeed, in terms of iron manufacturing in the past, the Weald has benefited from more research than any other such area in the country. What I attempt to achieve here is thus to synthesise a very diverse range of sites that have been excavated and reported over a very lengthy period of time and in many different styles, updating them where new information is available.

In terms of these sites, in many cases the existence of the occupation-period iron manufacturing there first came to light when their associated cinder heaps and fields were utilised to provide material to metal road surfaces for the region's poor quality trackways in the 19th and early 20th centuries. The ensuing antiquarian investigations by the likes of Lower and Rock then provided the first insight into a Roman industrial presence in the Weald which had been lost to living memory (even given a number of site names specifically reference their past, for example Cinderfield). These antiquarians developed relationships with the local workmen employed to dig the cinder heaps and fields such that as and when Roman material was found they were contacted to investigate (with money likely changing hands). These early investigations developed over time into the full suit of such activities employed by the modern field archaeologist, including full excavation, test pitting, field walking, walkover surveys and geophysical surveying, all backed up by modern scientific analysis and interpretation.

In terms of the evidence itself, each site I detail considers some or all of the following features: buildings (including substantial amounts of roof tile, often with a *Classis Britannica* stamp, see below), burials, iron manufacturing detritus (including charcoal), pottery, coins, glass and other associated small finds. The latter four are often found within the cinder heaps themselves. Dating throughout the period of investigation of the occupation-period Weald has largely relied on pottery and coins, *Classis Britannica* stamped tile and to a lesser extent glass and other aspects of material culture, with charcoal being used more recently. Where any of the latter have been used in association with pottery the dating is more reliable given the tradition of low material culture consumption in the region during the LIA and occupation, and the proliferation of long-lived handmade wares such as East Sussex Grog Tempered Ware (Lyne, 359).

3.1.2 Industrial Origins

Iron production began in the Weald in the LIA on the northern and southern fringes of the region and it is clear that the arriving Romans had detailed foreknowledge of the industrial potential of the area. This is evident in the Caesarian citation of '*iron in the maritime*' (The Conquest of Gaul, V.135) which Hodgkinson (2008, 28) argues references the Wealden iron industry, he believing the same of Greek geographer Strabo's comment about iron being an export from Britain (The Geography, IV.5).

Two recent finds of datable charcoal in an iron manufacturing context at Tablehurst Farm in Forest Row and Cullinghurst Wood near Hartfield could illustrate the LIA beginnings of the industry in the region, the latter potentially as old as the 8th century BC. In total 23 bloomery sites in Weald may actually date to the LIA. This incudes high profile ones such as Broadfields, Garden Hill, Crowhurst Park and Footlands Farm.

Things accelerated with the arrival of the Romans, again illustrating the high degree of their foreknowledge, with the Weald rapidly became a key centre of industrial activity. This was based on the availability of the siderite iron ore which sits within both the Wealden Clays and the Hastings Beds (particularly the Wadhurst Clays in the case of the latter), providing a ready source of raw material to facilitate an iron manufacturing industry which thrived until the middle of the 3rd century AD (a chronological pattern shared with the Medway Valley and its ragstone quarrying industry, see Section 4.3 below).

Mattingly (2006, 386) argues that the Weald was specifically very different in nature to both the north west and eastern Kentish economic regions during the occupation, largely because of the comparative lack of evident settlement and the heavy focus on industrial activity. Recent research by Harrington and Welch (2014, 109) has further helped refine our knowledge of the Wealden industrial experience during the occupation, substantiating earlier work summarised by Jones and Mattingly (1990, 193). They have identified that there were two very specific industrial regions in the Weald, each with evident iron working but on differing scales. The central region had some large sites but mainly featured localised iron working, perhaps on a seasonal basis by those already farming the land, and catered specifically for a regional demand. Harrington and Welch (2014, 109) argue that iron manufacturing here was aligned northwards towards London (Southwark in particular) and is associated with the main occupation-period roadways and pre-existing LIA trackways in Sussex. Principal sites included Broadfields, Great Cansiron and Oldlands.

However, it is in the eastern/ coastal region that massive industrial operations are visible at some sites in the archaeological record. These have long been widely linked with a state-presence (often in the form of the *Classis Britannica,* for example by Brodribb, 1979, 141, and Cleere and Crossley, 1995, 64), this being discussed in detail in Chapter 5. Sites here produced much of the iron for Britain and the near Continent until the industry's regional demise in the mid-3rd century AD. They catering for national and international demand, particularly with regard to the Roman military presence in the north of Britain. Principal sites in this eastern/ coastal region included Bardown, Beauport Park (the site of much of Brodribb's groundbreaking work), Chitcombe, Oaklands Park, Footlands Farm and Crowhurst Park.

Other industries also thrived in the Weald in addition to occupation-period iron manufacturing, often in some kind of association with it. In this regard Cleere and Crossley (1995, 81) highlight the timber and charcoal production industry essential to support the iron manufacturing, while the Weald is also well known for the manufacture of brick and tile, often featuring the stamp of the *Classis Britannica* regional navy as detailed above (Peacock, 1977, 236). Further, Lyne (1994,134) highlights pottery manufacturing, for example the Alice Holt industry on the borders of the western Weald. Peacock (1987, 61 also identified an additional extractive industry exploiting natural resources, this being the quern production site and its associated quarry at Lodsworth in West Sussex at the extreme western edge of the Weald, while other regional quarrying can be referenced with regard to materials such as Ashdown Sandstone from the lowest strata of the Hastings Beds. Meanwhile, Shaffrey and J.R. Allen (2014, 288) have also identified an important Roman whetstones industry based on the sandstones of the Wealden Clay in the north west Weald, though so far no associated quarry or production site has been found.

3.1.3 Site Selection and Grading

Here I have focused only on those sites where iron was being processed in some way and not those where the evidence points to mercantile activity only, for example the likely ports at Castle Croft and Bodiam, the peripheral small town at Arlington (at TQ 541 068), the peripheral villa site at Barcombe (at TQ 417 142) and, with regard to the latter, the adjacent small settlement at Bridge Farm (at TQ 433144).

Regarding the iron working sites, those in the occupied Weald came in a wide variety of shapes and sizes, ranging from the single furnace site at Pippingford Park to that of 10 ha at Beauport Park. Hodgkinson (2008, 1) says in this regard: *"They vary in size from sites where the estimated volume of waste is less than 10m^3 to a few where it is over 10,000m^3."*

He has helpfully developed a grading methodology based on the size of each site's waste material heap to help categorise the different sizes of iron making operations in the Weald during the occupation, these ranging from Grade 1 (smallest) to Grade 4 (largest). This is detailed in Table 3.1 and then utilised in the following review of the primary evidence sites.

Analogy can be drawn here with Peacock's (1982, 9) modes of production for the Roman pottery industry, where there are clearly parallels (see Section 2.4). For example Site Grade 1 would fit with the household production/ household industry modes, Site Grade 2 with individual workshops, Site Grade 3 with nucleated workshops and Site Grade 4 with manufactories. An extra layer of complication is added at some of the larger sites which have a *Classis Britannica* association as they would also potentially fit with the military and official production model (1982, 11).

Some 114 sites are now known where iron working took place in the Weald during the occupation (Hodgkinson, 2009, 31), though from Table 3.1 above it is clear that the larger sites dominated the total volume of iron produced, despite being much fewer in number. Also noteworthy is the fact that the three sites producing waste volumes of over 10,000m^3 are in the eastern/ coastal region, they being Beauport Park, Oaklands Park and Footlands Farm (see Section 3.1.4).

Regarding methodology, Cleere and Crossley's analysis of 40 of the key Wealden occupation-period iron manufacturing sites has been utilised by those studying such activity in the region for almost 40 years and has yet to be academically updated to take into account the significant amount of research that has occurred since. I do that here, reviewing the sites where some level of iron manufacturing is known to have been taken place to extrapolate the 37 that I determine to have been the most significant. The specific criteria used to determine the inclusion of the 37 in the list of primary evidence sites, referencing WIRG's own classification system, are as follows:

- Size 3 and 4 sites, based on Hodgkinson's waste-material heap size as outlined above.
- Sites featuring other significant economic activity, for example use as a port or it being a small town, in addition to iron manufacturing.
- Other sites featuring buildings, including those with *Classis Britannica* tile.
- Other sites featuring either significant quantities, or high quality examples, of elite pottery such as Samian ware.

Each selected site is considered in terms of the dates for iron manufacturing activity, specific site type, and finally the site history and known economic evidence (a pattern replicated in the later primary evidence site lists for the Folkestone region and the Medway Valley). Further, the sites are additionally classified as being either in the central or the eastern/ coastal region. Where *Classis Britannica*-stamped tiles are mentioned, all are Fabric 2 (see Section 3.6 for detail) unless otherwise specified. A final point is with regard to the 2008 High Weald coin hoard found at an occupation-period satellite iron working site near Bardown which is not detailed here as the location of its finding is publicly undisclosed. (it being considered in the discussion in Chapter 5).

The sites are listed below in alphabetical order.

Table 3.1: Romano-British ironworking sites in the Weald: waste heap grade data

Site Grade	Site Waste Volume	Total Volume	% of Sites	% of Total Volume
1	<=100m^3	2,437	48	1.6
2	101-1000m^3	15,200	32	10.2
3	1001-10,000m^3	66,000	16	44.4
4	>10,000m^3	65,000	4	43.7

Hodgkinson, 2008, modified by Elliott 2013.

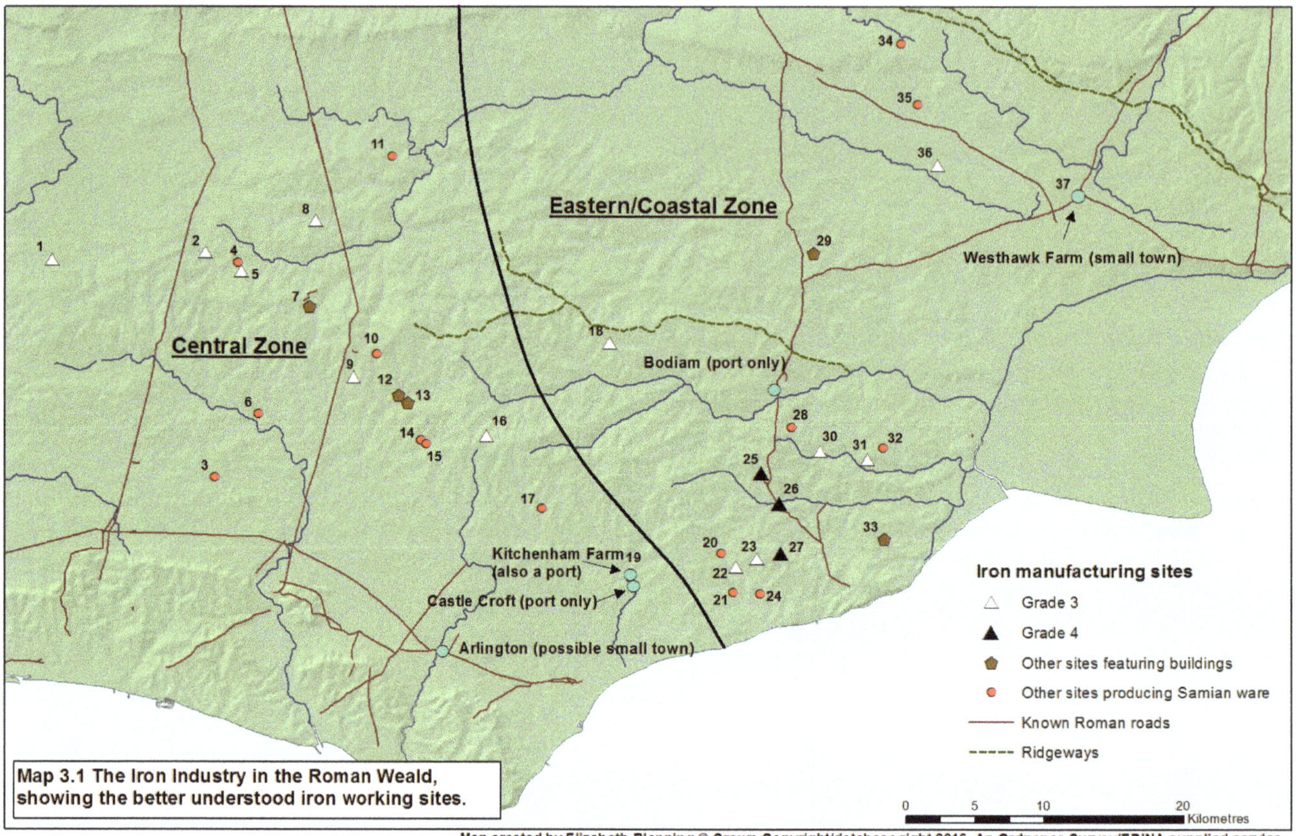

Map 3.1 The Iron Industry in the Roman Weald, showing the better understood iron working sites.

Map created by Elizabeth Blanning © Crown Copyright/database right 2016. An Ordnance Survey/EDINA supplied service.

Key	Site	Key	Site
1	Broadfields	20	Pepperingeye
2	Ridge Hill	21	Bynes Farm
3	Grassy Wood	22	Forewood
4	Standen	23	Crowhurst Park
5	Walesbeech	24	Upper Wilting Farm
6	Coalham	25	Footlands
7	Garden Hill	26	Oaklands Park
8	Great Cansiron	27	Beauport Park
9	Oldlands	28	Colliers Green
10	Newnham Park, Chillies Farm	29	Little Farningham
11	Oakenden	30	Chitcombe
12	Morphews	31	Ludley Farm
13	Howbourne Farm	32	Glossams Place
14	Pounsley	33	Icklesham
15	Cinderfield/Mill View Farm	34	Runhams Farm
16	Crawlsdown Wood	35	Coldharbour Farm
17	Blackman's Farm	36	Romden
18	Bardown	37	Westhawk Farm
19	Kitchenham Farm		

3.1.4 Site List

Central Region

Blackman's Farm (Grade 1, at TQ 6140 1720, number 17 on Map 3.1)

Date: Undated.

Type: Industrial site.

Site History and Economic Evidence: This site was discovered in 1977 by the Hailsham School Practical History Group (unpublished). It was then investigated by WIRG (Farebrother, 1978, 5), before being considered in detail by Cleere and Crossley (1995, 304) and Hodgkinson (1999, 71). A 20m by 15m spread of tap slag was found here along with charcoal and part of a clay tuyère (furnace nozzle), together with several samian ware sherds.

Broadfields (Grade 3, at TQ 258 353, number 1 on Map 3.1)

Date: 1st century BC through to 3rd century AD (based on pottery data).

Type: Industrial site.

Site History and Economic Evidence: John Gibson-Hill (1972, 25) excavated this site in the 1970s, it then being considered in detail by Cleere and Crossley (1995, 297). A key High Wealden site for occupation period iron production in the central region, activity here began in the LIA and continued until at least the 3rd century AD. Archaeological data includes a total of 36 large furnaces, with Cleere and Crossley (1995, 81) speculating that annual iron production would have been in the region of 50 tonnes.

Cinderfield/ Mill View Farm (Grade 2, at TQ 5290 2190, number 15 on Map 3.1)

Date: Undated.

Type: Industrial site.

Site History Economic Evidence: This site was excavated in the 1970s by WIRG under the leadership of Joseph Pettitt (1973, 13), then being considered in detail by Hodgkinson (1999, 70). A large concentration of tap slag has been found here across two fields and in a nearby stream, together with pottery including sherds of Samian ware (Pettitt, 1973, 13).

Coleham (Grade 1, at TQ 4070 2410, number 6 on Map 3.1)

Date: Undated.

Type: Industrial site.

Site History and Economic Evidence: Tebbutt (1978a, 405) led WIRG's excavations here in the 1970s, it then being considered in detail by Cleere and Crossley (1995, 299) and Hodgkinson (1999, 71). A scattering of bloomer slag was found in association with six sherds of Samian ware.

Crawlsdown Wood (Grade 3, at TQ 5735 2250, number 16 on Map 3.1)

Date: 1st through 4th century AD (based on pottery data).

Type: Industrial site.

Site History and Economic Evidence: This site was excavated by WIRG in the 1990s (Hodgkinson, 1997, 3), then being considered in detail by Hodgkinson (1999, 3). The site covers an area of 1 ha and features a number of banks of bloomery waste.

Garden Hill (Grade 1 at TQ 444 319, number 7 on Map 3.1)

Date: 1st century BC through to the 2nd century AD (based on pottery and glass data).

Type: Industrial site (and possible later small town).

Site History and Economic Evidence: This site was first located by Tebbutt during a trial excavation in 1968 (unpublished) with it becoming the subject of lengthy excavations from 1972 to 1978 by the Garden Hill Excavation Group led by J.R. Money (Money, Fulford and Eade, 1977, 339, and Money, 1979, 16). It has since been considered in detail by Cleere and Crossley (1995, 300) and Hodgkinson (1999, 71).

The 1970s excavations found extensive industrial and built infrastructure, including 1st century AD roasting and smelting furnaces, a forging hearth together with associated timber buildings, a structure with two verandahs, a 2nd century AD stone-built bath house and a timber building built on a stone base. Hodgkinson (2009, 28) argues that iron production here began in the LIA and continued until the early 2nd century AD, by which time the site was being eclipsed by the larger state-run iron production sites which had grown up in the eastern/ coastal region. He adds that iron production here was limited, but given the associated buildings it may have had some administration function for the central region.

Grassy Wood (Ungraded, at TQ 3753 1949, number 3 on Map 3.1)

Date: Undated.

Type: Industrial site.

Site History and Economic Evidence: This site was identified in 1993 by South East Archaeology Services

(SEAS) during a monitoring exercise for the construction of a new water main from Ditchling to Wivelsfield Green (unpublished), it then being investigated and recorded by WIRG the following year (Hodgkinson, 1994, 2). Pottery including Samian ware together with tegulae roof tile were found at the site in association with a concentration of tap slag, cinder and burnt clay.

Great Cansiron (Grade 3, at TQ 448 382, number 8 on Map 3.1)

Date: 1st through 3rd centuries AD (based on pottery and coin data).

Type: Industrial site.

Site History and Economic Evidence: The occupation iron-working provenance of this large site was first noted by Margary (1951, 100) when examining antique maps while tracking the route of the Roman road from London to Lewes which lies 2km away. It was then investigated by Tebbutt in the 1970s (1971, 11, 1972, 10, and 1979, 14). The site has since been considered in detail by Swift (1982, 20), Cleere and Crossley (1995, 299) and Hodgkinson (1999, 70). Finally, it was recently the subject of an extensive geophysical survey by Russel and Staveley (2012, 1).

Cleere and Crossley (1995, 981) speculate that iron industry operations here began under private control, with activity continuing at the site until the 3rd century. Most recently Russel and Staveley's survey has provided new insight into the location, revealing evidence to illustrate this was an extensive iron-working community. Of particular importance they say: *"...along the old stream frontage (at the site) there may be evidence for quays, suggesting links with the CLBR."*

This is remarkable for two reasons. Firstly, if further research confirms a *Classis Britannica* association then it would be the first such site in the central region (as a note of caution, none of the tile found here to date are *Classis Britannica*-stamped). Secondly, the nearest major waterway to Great Cansiron is the River Medway (accessed from a stream running through the site and where the 'quays' were found), far upriver from the currently known furthest upriver occupation-period site at Teston and thus (if proved) dramatically extending occupation-period activity in the river valley. Such a far upriver location is not uncommon in Kent, with Durham and Goormachtigh (2015, 174) commenting on how some emporia in the county are 'amazingly' so in their location.

Meanwhile, Great Cansiron is also associated with the ironstone quarries at nearby Tugmore Shaw. Hodgkinson (2013) says:

> *"The quarries here appear to be part of the Great Cansiron complex...large open-cast workings on the west side...seem to have predated smaller shaft mine pits which are profuse throughout the area suggesting that the larger quarry pits may be associated with the Roman works at Great Cansiron."*

Howbourne Farm (Grade 2, at TQ 516 249, number 13 on Map 3.1)

Date: 2nd century AD (based on pottery data).

Type: Industrial site.

Economic Evidence: One of the earlier identified occupation-period iron-working sites in the Weald, this location is first recorded by Straker (1931, 390) before being investigated by Tebbutt in 1972 (1973, 115). It was then considered in detail by Cleere and Crossley (1995, 299) and Hodgkinson (1999, 70). A bloomery has been found here in association with much occupation period pottery including a number of sherds of good quality Samian ware. Walling and window glass also indicate a building associated with the iron industry on the site.

Kitchenham Farm (Grade 2, at TQ 678 124, number 19 on Map 3.1)

Date: 1st and 2nd century AD (based on pottery, coin and tile data).

Type: Industrial site and port.

Site History and Economic Evidence: First recorded by WIRG (Cleere, 1969, 18) and then investigated by Tebbutt (1976, 324), this site near Ashburnham on the eastern edge of the Pevensey levels has been under excavation and investigation by HAARG since 2007 (Cornwell and Cornwell, 2008, 10, 2008, 1, and 2010, 16). During the occupation Kitchenham Farm was both a port supporting nearby iron working sites and also such a site in its own right. Data supporting the port interpretation includes the location of *in situ* timbers found during the 1970s investigations which were interpreted as the remains of an occupation-period jetty or landing stage. At the time of its operation such a port would have made use of tidal access to Pevensey Bay to facilitate maritime commerce.

Of great interest to this research, the recent investigations by HAARG at Kitchenham Farm have located 31 *Classis Britannica*-stamped tiles, of which 29 are of the Fabric 1 type originating from Boulogne (Cornwell and Cornwell, 2010, 16, see Section 3.6 for definition). These are the first *Classis Britannica* tiles located in the central Weald. Also found during the 1970s and more recent investigations were a large quantity of other types of building material, Samian ware pottery (the earliest pieces being Flavian), and over 1,000 occupation-period coins. Most recently a resistivity survey by HAARG as part of the ongoing investigation here has identified a number of features which indicate the location of specific buildings.

Morphews (Grade 2, at TQ 509 255, number 12 on Map 3.1)

Date: 1st century AD (based on pottery data).

Type: Industrial site.

Site History Economic Evidence: Another site first recorded by Straker (1931, 389), the large slag heap here was much used locally for road metalling. It was considered in detail by Cleere and Crossley (1995, 297) and Hodgkinson (1999, 70). Key finds here include a large occupation period bloomer together with hypocaust tile indicating a building nearby.

Newnham Park, Chillies Farm (Grade 1, at TQ 4935 2843, number 10 on Map 3.1)

Date: Undated.

Type: Industrial site.

Site History and Economic Evidence: Tebbutt led excavations here in the late 1970s and early 1980s (1978b, 7, and 1981b, 62). It was then considered in detail by Cleere and Crossley (1995, 296) and Hodgkinson (1999, 71). A thick 50cm-deep layer of tap slag and furnace lining material has been identified here in association with a number of sherds of Samian ware.

Oakenden (Grade 2, at TQ 504 428, number 11 on Map 3.1)

Date: 2nd century AD (based on pottery data).

Type: Industrial site.

Site History and Economic Evidence: A large 30m by 80m cinder field has been found here including tap slag and charcoal, together with pottery on the surface which includes Samian ware sherds. It was recorded by Cleere (1974, 197), and then considered by Cleere and Crossley (1995, 380) and Hodgkinson (1999, 70).

Oldlands (Grade 3, at TQ 476 268, number 9 on Map 3.1)

Date: 1st through 4th centuries AD (based on pottery, coin and glass data).

Type: Industrial site.

Site History and Economic Evidence: This was the first occupation-period iron-working site identified in Sussex, being discovered by the Rev. Edward Turner (Rector of Maresfield) in 1844. Lower (1849, 171) recordered that Turner found Roman pottery in a cinder heap piled beside a local roadway ready to be used for repairs. Inquiring locally of the origins of the cinders he was directed to the site of the Roman iron-works which he found being uncovered by workmen digging out the cinder heap. Lower (1849, 171) added that Turner was also made aware that several inhumation burials of occupation-period date had also been found in living memory at the site, the graves being cut deeply into surrounding cinder fields which had fallen out of use as the occupation progressed. Lower (1849, 171) further detailed that Turner had found a building at the site measuring 9.1m by 3.7m. The site was later investigated by Straker (1931, 395) and Tebbutt (Tebbutt and Tebbutt, 1982, 12). It was then considered in detail by Dalton (1983, 34), Cleere and Crossley (1995, 302) and Hodgkinson (1999, 70).

Data from these investigations and subsequent analysis has revealed that iron manufacturing began here in the 1st century AD and continued until the 4th century AD. As with Great Cansiron, Cleere and Crossley (1995, 74) reasonably speculate that iron manufacturing may have been initiated under private ownership, with ore provided from the opencast mining of nearby deposits. Material culture from the site includes coins ranging in date from Nero in the 1st century AD to Diocletian in the late 3rd/ early 4th AD, Samian ware and much other pottery, glass, lead used in a building context and a stylus.

Pounsley (Grade 2, at TQ 525 222, number 14 on Map 3.1)

Date: 2nd century AD (based on pottery data).

Type: Industrial site.

Site History and Economic Evidence: This site was first recorded in detail by Cleere (1974, 198), before being considered by Cleere and Crossley (1995, 300) and Hodgkinson (1999, 70). Cleere's investigations revealed a slagheap on the banks of a stream which included a wide variety of occupation-period pottery including a number of sherds of Samian ware.

Ridge Hill (Grade 3, at TQ 369 359, number 2 on Map 3.1)

Date: 1st through 4th centuries AD (based on pottery data).

Type: Industrial site.

Site History and Economic Evidence: Excavations began here in 1927 under Straker (1928, 183) after local farmers found Roman pottery in an exposed section of 150m by 60m cinder heap on a stream bank (this being dated by S.E. Winbolt in a letter to Straker dated 19th December 1927, Straker, 1931, 235). It was later considered in detail by Margary (1933, 177), Cleere and Crossley (1995, 298) and Hodgkinson (1999, 70). Data from the investigations and subsequent analysis indicates iron production here began in the 1st century AD and continued through to the 4th century AD. One of the most northerly of the central group of significant sites in the Weald, Cleere and Crossley (1995, 61) argue that it was originally set up to exploit ore deposits located during occupation-period regional road building, with its market outlet being northwards to London.

Standen (Grade 2, at TQ 3920 3510, number 4 on Map 3.1)

Date: 2nd century AD (based on pottery data).

Type: Industrial site.

Site History and Economic Evidence: First recorded by Straker (1931, 239), this site was later excavated by he and R.T. Mason (Straker and Mason, 1939, 153). It was considered in detail by Cleere and Crossley (1995, 380) and Hodgkinson (1999, 71). The site features a large 50m slagheap in which some 14 sherds of pottery have been recovered including Samian ware.

Walesbeech (Grade 3, at TQ 395 345, number 5 on Map 3.1)

Date: 1st through 3rd centuries AD (based on pottery data).

Type: Industrial site

Site History and Economic Evidence: The slag heap for this site was first observed by Straker who went on to excavate it and the surrounding area with Margary (Straker, 1931, 239). It was later considered in detail by Cleere and Crossley (1995, 298) and Hodgkinson (1999, 70). The investigations here revealed that industrial activity began in the 1st century AD and continued until the 3rd.

Eastern/ Coastal Region

Bardown (Grade 3, centred at TQ 6632 2928, number 18 on Map 3.1)

Date: 1st through 3rd centuries AD (based on pottery, coin and tile data).

Type: Industrial site.

Site History and Economic Evidence: This important site on the south bank of the River Limden near Ticehurst was discovered in February 1909 by a Mrs Odell and Mr Eden Dickson (Straker, 1931, 296), being initially dated through the analysis of Roman pottery found at the location (Haverfield, 1916, 195). Initial investigations by Straker (1931, 298) included the finding locally of a small clay lamp and an occupation-period beakless anvil. The site was fully excavated by Cleere in the 1960s who recorded his activities in *'The Romano-British Industrial Site at Bardown'* which was published by the SAS in 1970. In particular he focused on the massive occupation-period refuse tip at the site which spreads over 100m, comprising tap slag, cinder, furnace debris and domestic waste. Also found were the occupation-period roasting hearths, forges and a charcoal-burning hearth which helped identify the principal iron-working areas. More recently, in 2006 during a metal detectoring exercise the IHRG found a fine quality medallion dating to the reign of Antoninus Pius (Emperor AD 138 to AD 161), while in 2009 HAARG located two circular enclosures near the site exit. Most recently a geophysical survey by David Staveley has failed to find any sign of a settlement at Bardown outside the context of the known industrial activity (Hodgkinson, 2012, 1). The site remains a live project for the IHRG, and has been considered in detail by Cleere and Crossley (1995, 303) and Hodgkinson (1999, 70).

Bardown is a large northern outlier of the eastern/ coastal iron- manufacturing region in the Weald, with data from the presence of *Classis Britannica*-stamped tile (28 in total) at the site and its southerly transport links to the coast being used to determine its home here rather than in the central region. Widely associated with the regional navy (Cleere and Crossley, 1995, 70), around 50 tonnes of iron would have been produced here each year, though after iron-making activity later moved to nearby satellite sites in the early 3rd century AD the original industrial range was re-used as a domestic rubbish dump. Describing the site when iron production was at its height, Cleere and Crossley (1995, 75) say: *"Bardown represents the military ironworks admirably. The only substantial building on the site was a sturdy timber-framed barrack block of a standard type which would have housed some 40 men."*

A bath house may also have existed here based on fragments of box flue tile found in the area, though the building has yet to be located.

Beauport Park (Grade 4, at TQ 786 140, number 27 on Map 3.1)

Date: 1st through 3rd centuries AD (based on pottery, coin, tile and glass data).

Type: Industrial site.

Site History and Economic Evidence: This site was first identified by its very large 1 ha slag-heap in 1862 by the Rev. S. Arnott (1862, 138), then being investigated and recorded by Rock (1879, 168). It then became a local source of material for metalling roads and building in the late 19th century, with the County Highways Supervisor quarrying away up to 30,000m^3 of the waste over a 10 year period. Beauport Park was then recorded by Straker (1931, 330), before being excavated in detail by Brodribb and Cleere in the 1970s and 1980s (Brodribb, 1979, 139, and Brodribb *et al*, 1988, 232). It was further investigated in the early 1990s by WIRG (Hodgkinson, 1991, 2) before being considered in detail by Cleere and Crossley (1995, 295) and Hodgkinson (1999, 70). Next, it was the subject of a Time Team episode which was broadcast as episode seven of series six in 1999. Most recently Staveley (2013) has argued that the Rochester-Wealden road terminated at Beauport Park rather than Hastings as previously argued by Margary (1967, 46).

Beauport Park was the second (and largest) of the principal iron ore extraction sites in the Weald during the occupation. Iron production began here in the 1st century AD and continued until the 3rd century AD (Cleere and Crossley, 1995, 70). The site is the largest of the Wealden iron production sites from the occupation period, being up to 10ha in size in total including the original but now quarried away 1 ha slag heap. The latter was well detailed by Straker (1931, 330), he saying that the excavations by workmen left exposed sections of the spoil which revealed it to have been laid down in 25cm layers, each of which was divided into four sub-layers. These latter comprised a sub-layer of charcoal, then one of burnt earth, then one of iron ore waste and finally one of burnt clay.

Based on its very large size and central location in the eastern/ coastal region, Cleere and Crossley (1995, 70) believe that a case can be made for Beauport Park actually being the headquarters of any state-run iron production in the region during the occupation, a view supported by Hodgkinson (2009, 33). Staveley's (2013) view that the Rochester-Wealden road terminated here certainly supports this view. It is further supported by tile evidence, given the very high proportion of Fabric 2-style Wealden *Classis Britannica* tiles found here (51 complete out of 1,600 in total featuring the regional navy stamp), this at the very least demonstrating a strong association with the regional fleet. The tiles have been found in the context of a high quality six-room military-style bath house with a floor area of 114m^2 (Brodribb, 1979, 141), although there is evidence in the form of uninvestigated foundations that settlement here was far more extensive (Cleere and Crossley, 1995, 295). The state- association hypothesis is further bolstered by two additional pieces of epigraphy, the first in the form of a reference on the bath house stonework entrance to a *vilicus* official who ran the site (Brodribb *et al*, 1988, 261), and the second on a tile comb featuring the stamp of the *Classis Britannica* (Brodribb *et al*, 1988, 269, see Chapter 6 for discussion on both). Other material culture finds at the site have included coins, pottery and glass dating from the 1st through 3rd centuries AD, a bronze ring and a bronze *ligula*.

Bynes Farm (Grade 2, at TQ 752 111, number 21 on Map 3.1)

Date: 1st and 2nd century AD (based on pottery data).

Type: Industrial site.

Site History and Economic Evidence: First recorded by Straker (1931, 358), this site was excavated in 1949 by B.H. Lucas (1950, 49) and then appraised by Hodgkinson (1999, 70). Identified as a satellite site for Crowhurst Park, the excavations in 1949 revealed a large amount of 1st and 2nd century AD pottery including Samian ware within the extensive slag heap.

Chitcombe (Grade 3, at TQ 814 214, number 30 on ap 3.1)

Date: 1st and 2nd century AD (based on pottery data).

Type: Industrial site.

Site History and Economic Evidence: This extensive site was first recorded by Rock (1879, 175), then Straker (1931, 345) and Austen (1947, 84). It was then investigated by WIRG (1988, 2) before being considered in detail by Cleere and Crossley (1995, 296) and Hodgkinson (1999, 70). The scale of iron production here is indicated by the enormous slag and refuse dump, with Cleere and Crossley (1995, 81) estimating this to total just under 10,000m^3 in volume. They say this weighed 30,000 tonnes, leading them to speculate that total iron production would have been 70 tonnes annually. Tile, though not of *Classis Britannica* provenance, has also been located here, and Rock (1879, 175) noted at the time of his initial investigation that occupation-period buildings were still standing on the site though these have not been found since.

Coldharbour Farm (Grade 1, at TQ 8840 4660, number 35 on Map 3.1)

Date: 2nd century AD, based on pottery data.

Type: Industrial site.

Site History and Economic Evidence: This site was recorded by Kaminski (1995, 129), it being close to the nearby London-Brighton Roman Road. A scatter of bloomer slag was found here in association with a cremation burial group (Kaminski, 1995, 129), within which was recovered a Samian ware dish.

Colliers Green (Grade 3, at TQ 7930 2310, number 28 on Map 3.1)

Date: 2nd century, based on pottery data.

Type: Industrial site.

Site History and Economic Evidence: First recorded by Straker (1931, 319), this site was investigated by Jones (1981, 69) and then considered in detail by Hodgkinson (1999, 70). A large concentration of slag was found here in association with pottery sherds including Samian ware.

Crowhurst Park (Grade 3, at TQ 769 136, number 23 on Map 3.1)

Date: 1st century BC through to 3rd century AD (based on pottery data).

Type: Industrial site.

Site History and Economic Evidence: Recorded by Straker (1931, 353), this site was excavated in the later

1930s by Piggott (1937, 231) and Straker and Lucas (1938, 224). It was then considered in detail by Beswick (1991, 246), Cleere and Crossley (1995, 297) and Hodgkinson (1999, 70).

Iron production here may have begun in the LIA, and was definitely underway by the later 1st century AD, continuing until the 3rd century AD. Cleere and Crossley (1995, 81) believe that the total iron production here, based on the amount of slag still in existence today, would have amounted to some 50 tonnes annually.

Footlands Farm (Grade 4, at TQ 772 198, number 25 on Map 3.1)

Date: 1st century BC through to the 4th century AD (based on pottery and coin data).

Type: Industrial site.

Site History Economic Evidence: This site was discovered in 1924, being excavated the following year by the SAS (unpublished at the time), the results then being recorded by Straker (1931, 327) who noted the large amounts of LIA and occupation-period pottery visible. Further pottery finds at the site in the 1940s were recorded by Chown (1947, 148), with it then being considered in detail by Cleere and Crossley (1995, 303) and Hodgkinson (1987, 25, 1988, 231, and 1999, 70). Most recently the site has been the subject of a geophysical survey by HAARG (Cornwell, 2013, 1) which has revealed what appears to be an extensive roadside settlement at the site. More investigation with regard to the latter is planned though has yet to take place.

Footlands Farm is one of the largest iron manufacturing sites in the Weald based on data from surviving waste material (spread intermittently over 2ha), with iron production beginning in the 1st century AD and continuing well into the 4th (though later on a much more localized basis). Cleere and Crossley (1995, 62) hypothesise that iron manufacturing activity here could actually have begun in the LIA, as above with Crowhurst Park.

Forewood (Grade 3, at TQ 754 130, number 22 on Map 3.1)

Date: 1st century AD (based on pottery data).

Type: Industrial site.

Site History and Economic Evidence: This site was first recorded by Straker (1931, 351) and Smythe (1937, 197), then being excavated by WIRG in 1991 (Hodgkinson, 1992, 8) and 1993 (Hodgkinson, 1993, 2). It has since been considered in detail by Cleere and Crossley (1995, 303) and Hodgkinson (1999, 70). Forewood has been interpreted as one of the larger satellite sites for Crowhurst Park, with evidence for iron-working activity including a spread of over 100m of slag and an unworked iron bloom.

Glossams Place (Grade 2, at TQ 8590 2160, number 32 on Map 3.1)

Date: Undated.

Type: Industrial site.

Site History and Economic Evidence: The finding of a 20m spread of occupation-period bloomery slag was recorded here by Hodgkinson (1993, 2), it then being considered in detail by Cleere and Crossley (1995, 380) and Hodgkinson (1999, 70). Within the waste material were found a number of sherds of Samian ware pottery.

Icklesham (Grade 2, at TQ 86 15, number 33 on Map 3.1)

Date: 1st and 2nd century AD (based on pottery and coin data).

Type: Industrial site.

Site History and Economic Evidence: Noted by Straker though with no other details given (1931, 340), this site at Telegraph Mill, East Sussex was field walked by WIRG in the early 2000s. This exercise located a platform of slag detritus which was determined to have a volume of 600m^3 (Hodgkinson, 2007, 5). The bases of six shaft furnaces have also been found here, together with much bloomer slag and occupation-period material culture including Hadrianic coins. Tile has also been found, though none to date associated with the *Classis Britannica.*

Little Farningham (Grade 1, at TQ 809 358, number 29 on Map 3.1)

Date: 1st through 3rd centuries AD (based on pottery and tile data).

Type: Industrial site.

Site History and Economic Evidence: This site, located near Cranbrook, was first identified in the 1950s by site owner George Luck who recognised brick and tile being ploughed out of the ground as Roman, the location then being excavated by the Cranbrook and Sissinghurst Local History Society under the supervision of excavation secretary Cecily Lebon in the later 1950s and 1960s. Her papers on the investigation are held in Cranbrook Museum (Lebon, 1957, 224, 1958, xivii, and 1961, xiviii). It was then considered by Cleere and Crossley (1995, 380) before the site was revisited by Neil Aldridge with the support of the Kent Archaeological Society (KAS) in 1999 and 2000 (Aldridge, 2001, 135).

Little Farningham has been identified as an administrative site which also manufactured iron, based on the presence of occupation-period buildings and numerous stamped *Classis Britannica* tiles (51 in total, many of them in mint condition, Brodribb, 1970, 25) at such a small iron working site (Cleere and Crossley, 1995, 297). Of particular interest, a worked iron billet was found inside one of the buildings featuring stamped tile, together with evidence of a hypocaust system (Woodcock, 1998, 180). Pottery has also been found, dating from the 1st through 3rd centuries AD.

Ludley Farm (Burnthouse Wood, Grade 3, at TQ 848 208, number 31 on Map 3.1)

Date: 1st and 2nd century AD (based on pottery and coin data).

Type: Industrial site.

Site History and Economic Evidence: Located near Beckley, trial excavations here in the early 1970s revealed large amounts of pottery within a large 100m by 50m slag and refuse bank which had been much used for local road metalling (Scott, 1972, 29, and Botting, 73, 111). It was considered in detail by Cleere and Crossley (1995, 295) and Hodgkinson (1999, 70). Pottery from the site included Samian ware, with a few occupation period coins also being found.

Oaklands Park (Grade 4, at TQ 785 176, number 26 on Map 3.1)

Date: 1st through 3rd centuries AD (based on pottery and coin data).

Type: Industrial site.

Site History and Economic Evidence: This site was first identified as a Roman iron working location in the 1840s when Roman coins were found on land owned by Hercules Sharp (Lower, 1849, 174). Straker (1931, 329) reports that the extensive slag heap was heavily utilised in the 19th century as a source for metalling local roadways. After consideration by Cleere and Crossley (1995, 305) Oaklands Park was then investigated in the late 1990s and early 2000s by WIRG and HAARG (Hodgkinson, 2001, 3). Most recently a geophysical survey of the whole site by David Staveley was followed up by an extensive test pitting exercise by IHRG (Staveley, 2014) when a total of six large trenches were dug. The aim of the latter was to find *Classis Britannica* tiles to prove a link between the site and the regional fleet. None were found however, with Staveley noting a comparative lack of any tiles even though a number of buildings including the smithy were found.

From these investigations and associated data and analysis it is evident that that iron production took place at this large, eastern/ coastal site from the 1st century AD through to the 3rd century. Cleere and Crossley (1995, 91) say that the 20,000m^3 of slag still in existence, weighing some 60,000 tonnes, indicates a peak iron production rate of 140 tonnes per year. In that regard it can be considered the second largest Wealden iron production site after Beauport Park.

Pepperingeye (Grade 2, at TQ 743 140, number 20 on Map 3.1)

Date: 1st century AD (based on pottery data).

Type: Industrial site.

Site History and Economic Evidence: Located near Battle, this site was first recorded by Straker (1931, 351). It was then considered in detail by Cleere and Crossley (1995, 295), who thought it might be a satellite site for nearby Crowhurst Park, and Hodgkinson (1999, 70). Featuring a 1m layer of slag, Samian ware pottery has also been found here.

Romden (Grade 3, at TQ 8985 4220, number 36 on Map 3.1)

Date: 1st to 2nd centuries AD (based on pottery and coin data).

Type: Industrial site.

Site History and Economic Evidence: The Roman provenance of this site, on the extreme north eastern edge of the eastern/ coastal region, was first indicated by the finding of a 2nd century AD coin and possible associated burial in 1856 (unrecorded at the time). At the beginning of the 20th century this prompted Basil Worsfold, the owner of Romden Castle, to gather together some local antiquarians and carry out an investigation on his estate. This located significant amounts of iron slag and occupation-period pottery, though again at the time this went unreported. Both events were finally published by Worsfold in *Archaeologia Cantiana* (1931, 82). Field walking in 1994 by WIRG confirmed the Roman provenance of Worsfold's site, finding large amounts of additional pottery together with more iron slag (Aldridge, 1996, 16). Most recently, further field walking by WIRG in 2008 found 51 occupation-period pottery fragments including sherds of Samian ware. Also located was an extensive 200m by 150m cinder field featuring bloomery slag, furnace lining and roasted ore (Aldridge, 2009, 9), making it the largest Wealden Kent-based occupation period iron working site currently known.

Runhams Farm (Grade 1, at TQ 8720 5100, number 34 on Map 3.1)

Date: 1st to 2nd centuries AD (based on pottery data).

Type: Industrial site.

Site History and Economic Evidence: This site was first excavated by the Mid-Kent Training School in 1980, this being recorded by Tebbutt (1981b, 20). It was then investigated in detail over a lengthy period of time by KARU from 1978 to 1986 (Philp, 1994, 10). Runhams farm was then considered in detail by Hodgkinson (1999, 70). In terms of material culture, the remains of three bloomer furnaces have been found here together with a number of Samian ware sherds at what has been interpreted as a Roman-British farmstead featuring a limited amount of iron manufacturing.

Westhawk Farm (Grade 1, at TQ 9996 4000, number 37 on Map 3.1)

Date: 1st through 3rd centuries AD (based on pottery and coin data).

Type: Roadside settlement/ nucleated settlement and industrial site.

Site History and Economic Evidence: This well recorded small town, sited at the convergence of two major regional Roman roads, lies near modern day Ashford at the extreme eastern edge of the eastern/ coastal region. It was excavated in 1998 and 1999 by Oxford Archaeology, revealing a settlement of over 18 ha (Booth, Bingham and Lawrence, 2008, 272). WIRG recently rated it as a Grade 1 iron manufacturing site based on the identification of two small but intensive areas of iron working which have been associated with roadside trades such as shoeing horses or repairing vehicles (Paynter, 2007, 15). Booth (2001, 3) says that activity across the whole of the excavated area of the site declined dramatically by the mid-3rd century AD.

Upper Wilting Farm (Grade 2, at TQ 7710 1100, number 24 on Map 3.1)

Date: Undated.

Type: Industrial site.

Site History and Economic Evidence: This new site was excavated by Oxford Archaeology and HAARG between 2013 and 2014, though the results are currently unpublished. A 55m by 20m spread of slag alongside the remains of 14 bloomery furnaces have been found here, together with a number of Samian ware sherds.

3.2 Settlement in the Occupation Period Weald

There was comparatively little settlement in the Weald during the occupation, with Cunliffe (1988b, 84) explaining that: *"In the whole of the coastal zone between the Roman towns of Chichester and Canterbury, no (major) urban centre emerged."*

Modern research is now beginning to shed light on the limited amount that did exist there. For example HAARG have discovered the possible roadside settlement at the major eastern/ coastal iron working site at Footlands Farm mentioned above. Such a settlement would be analogous with that of the small town of *Ariconium* in the Forest of Dean where a similar roadside settlement developed in association with local iron working (Jackson, 2012, 195). A lively debate also exists about whether the sites at Garden Hill and Bardown were actually small towns, though such a hypothesis is to date unproven. Recent geophysical surveying work at Bardown certainly seems to now rule that site out as a settlement. Meanwhile, a final small town candidate with direct links to the Roman Wealden iron industry is found at Arlington in East Sussex on the periphery of the region, a road-side settlement sitting along the Greensand Way which linked Barcombe and Hardham at the place where this road crossed the Cuckmere River. This small town interpretation is again based on data from recent geophysical surveying and excavations by the Brighton and Hove Archaeological Society which have found roadside ditches, a large assemblage of occupation-period pottery (including extensive quantities of Samian ware), glass and potential structures including one constructed of masonry (Chuter, Washington and Corbett, 2008). Analogously this is a very similar situation to that of the Westhawk Farm small town near Ashford in Kent on the eastern edge of the Weald.

Even in the best case scenario however, it is clear that only a limited amount of occupation-period settlement is visible in the data from this extensive region, with most being in the context of industrial activity. Further, such settlement as there is evident only through to the middle of the 3rd century AD as detailed below in Section 3.4, after which both industry and its associated settlement begin to significantly diminish.

Given this limited settlement, and archaeological data from the Medway Valley where there appears to be little or no Roman presence further upriver than Teston from where the river meanders towards the Weald, I believe a good case can actually be made that access to the industrial region of the Weald was specifically restricted by the state during the occupation (see discussion in Chapter 5 in that regard). Cleere and Crossley (1995, 68) certainly viewed it as a single and homogenous unit, they arguing that this tradition continued into the post-Roman period when the name *wald* first appears as explained above, initially in association with the Saxon Shore fort at Pevensey (Roman *Anderida*, the initial Germanic name for the region thus being *Andredsweald*, Everitt, 1986, 53). It is instructive in this regard that there were clearly so few sites of cultural significance in the centre of the Weald that it was named after a site on its coastal periphery.

3.3 Transport Infrastructure in the Weald during the Roman Occupation

In the pre-modern era the Weald, given its heavy clay soils and wooded nature, was long associated with transport inaccessibility. Defoe (1724, 196) for example, in his tour of Britain in the early 18th century, said that it might take up to three years for a load of Wealden timber destined for the naval dockyards at Chatham to make the journey. Things were clearly better during the occupation however when a number of road-based and maritime options were available for transport purposes across the region, all designed to facilitate industrial activity there.

In terms of the former, the most important was the north-south Roman road which originated in Rochester where it branched off from Watling Street, then heading south through modern Maidstone before ultimately terminating in the environs of the huge Beauport Park iron manufacturing site to the immediate north of Hastings (having passed other significant iron working sites including Footlands Farm and Oaklands Park). Following the route of the modern A229 for much of its length (Vincent, 2007, 42), it was originally thought to actually go on to the south coast town but recent work by Staveley (2013) now shows this not to be the case. This roadway, identified as the principal land link between the north Kent coast and the Weald by Margary (1967, 272), allowed official communications to flow both ways and may also have provided access to the ragstone quarrying industry in the Medway Valley (see discussion in Chapter 5).

Margary (1967, 34) explained that access to the region (particularly the central area) was also provided by the radial Roman road network running south from London through to the south coast, particularly the roads to the likely port locations at Lewes (see above), Portslade and Chichester. This network was then linked west to east by other Roman roads running along the line of the escarpment foot of the South Downs and along the Sussex Coastal Plain. Margary (1967, 34) further detailed the existence of a network of Iron Age trackways across the Weald, for example those following the ridgeways which crossed the centre of the region. Harrington and Welch (2014, 109) add that most of the central region roadways and trackways, through a variety of types of connectivity, terminated at Southwark to the immediate south of *Londinium* (as does the Rochester-Wealden road via Watling Street). This location, principally catering for London, was a regional centre for a variety of types of metal working and craft during the occupation (Hodgkinson, 1994, 34) as identified from data provided by significant amounts of material culture and metal working detritus, found particularly at the Courage Brewery site there (Brigham *et al.* 1995, 1). This is directly supported by Cowan *et al* (2009, 106) in their definitive appreciation of Roman Southwark which, bringing together 41 previously unpublished excavation reports from 1973 to 1991, highlights over 60 iron-smithing and bronze working hearths used to make a range of metal goods including nails, tools and fittings.

All of these roadways and trackways, across the whole region, were together a principal factor in the location of the industrial sites in the Weald, with Cleere and Crossley (1995, 61) explaining that all such locations (whether in the central region or the eastern/ coastal region) were within 3.5km of such a roadway or trackway. Staveley (2013) actually believes that the Rochester - Wealden road and the radial routes from London were built specifically to open up the region to provide access to the industrial hinterlands between them, thus enabling the larger industrial sites to be built.

In terms of maritime transport, the occupied Weald was then as now well served by rivers providing access from the interior to the coast, for example in the west the Rivers Waller's Haven and Ouse giving access to the central region and to the east the Rivers Brede and Rother facilitating activity in the eastern/ coastal region. Cleere and Crossley (1995, 62) say that as the Wealden iron industry grew, specific ports were built on these rivers to facilitate the transportation of the pig iron being produced. The principal ports serving the central, more localised region to enable the transport of goods to the coast and onwards (as opposed to northwards along the road and trackways) would have been:

- Kitchenham Farm near Ashburnham, on the Ash Bourne (a tributary of the Waller's Haven) which would have had tidal access to Pevensey Bay during the occupation given that the coast line to the east of Pevensey was much further inland (at least tidally) prior to subsequent silting and land reclamation in the Pevensey Levels (Cornwell and Cornwell, 2008, 10, 2008, 1, and 2010, 16).
- Castle Croft on the Wallers Haven itself, the site of an unusual earthwork where Roman tile and coins have also been found and which would, during the occupation, also have had tidal access to Pevensey Bay (Cornwell *et al*, 2007, 3).

These two sites may have actually been linked in some way given their close proximity, for example one being the site of the actual port and the other the location of warehousing.

Additionally, recent data provided by geophysical survey work carried out by Russel and Staveley (2012, 1) has indicated the possible presence of quays along an old stream frontage at the central Wealden site at Great Cansiron, as noted in Section 3.1.4. Well away from the coast, this would have provided riverine access to the nearby River Medway and thus a route north to the ragstone quarries of the Medway Valley and on to the north Kent coast.

Meanwhile, for the eastern/ coastal region Cleere and Crossley (1995, 61) discuss two river port locations to facilitate coastal access for the much larger-scale state industry present there during the occupation. These are:

- An as yet unidentified site on the River Brede, the presence of which is indicated by the close proximity of the early iron manufacturing sites (though noting the highly speculative nature of their suggestion here).
- An extensive facility at Bodiam, where the Rochester-Wealden Roman road crossed the River Rother and where *Classis Britannica* tile has been found (Adler, 2013, 62). Cunliffe (1988b, 84) speculated that Bodiam may have actually been the principal occupation-period inland terminal facilitating the wider riverine transport of goods to the coast in this eastern/ coastal region, while Cleere and Crossley (1995, 64) highlight the fact that the location is equidistant from all of the later

regional major iron production facilities. In this regard it may have succeeded their earlier speculated port on the Brede.

Cunliffe (1988b, 84) added a further potential eastern/ coastal port location:

- A site on the northern edge of Romney Marsh near Lympne, site of the later Saxon Shore fort in whose walls re-used tiles from earlier structures have been found. He referenced the antiquarian discovery of large amounts of Roman period 'occupation debris' at Dymchurch as suggesting the potential location for such a port (also see Haverfield and Wheeler, 1932, 55).

Maritime access to these eastern/ coastal port sites would have been markedly easier during the occupation compared to the present day due to the subsequent silting up and land reclamation of the extensive area now covered by Romney Marsh, paralleling the experience to the west around Pevensey. At that time, the area now covered by Romney Marsh would have included an extensive area of sheltered water, thus being ideal for maritime trade. Cunliffe (1988b, 83) explained in his work with the Romney Marsh Research Trust that the shelter would have been provided by a coastal shingle barrier formed by long shore drift, and that it was into this body of water protected by the barrier that the drainage channels of the Brede and Rother would have flowed.

An excellent example of regional market integration (within both the Imperial and provincial economies), one should reflect here that it was of course not just iron being transported through this maritime trade network, but other materials also. For example Allen and Fulford (1999, 179) highlight the fact that locally quarried Ashdown Sandstone was also being exported, for use in the walls of the Saxon Shore fort at Brancaster in East Anglia. They speculate that this trade may have used its own bespoke ports built in the area between Bexhill and Fairlight.

A final point here is to consider the destinations of travel of the exported goods, particularly from the eastern/ coastal Wealden ports. The models of Morris (2010, 1) and Evans (2013, 433) are particularly useful in this regard, with the former's Southern North Sea and Eastern Channel System providing a ready framework to track goods travelling up the east coast and across the North Sea and the Channel to the Continent. Meanwhile the latter's eastern coast route provides an even more comfortable fit for Wealden goods travelling northwards, with the iron industry being particularly important supporting the extensive military presence in the north. As is demonstrated above in Section 2.5 however, trade would also have been important up the west coast where, once again, the military presence in the north and west would have been the principal beneficiaries.

3.4 The Iron Industry in the Weald during the Roman Occupation

The Weald was one of the three principal iron-producing areas in Britain during the occupation, the other two being the Forest of Dean and the East Midlands (Mattingly, 2006, 509). Both of the latter superseded the Weald in terms of importance after the middle of the 3rd century AD. Iron was also produced on a lesser scale elsewhere in Britain, for example in East Yorkshire (the industry there having its roots in the LIA, Halkon, 2011, 148), Exmoor in the south west and, later in the occupation, the Thames Valley.

Easily accessible raw materials were at the heart of the location of the Wealden iron industry, for example the region's siderite iron ore which had an average iron ore content of 40% (Jones and Mattingly, 1990, 192). The heavily wooded Weald was also a ready source of the large amounts of timber needed to produce the vast quantities of charcoal required for the iron manufacturing process, with Oak, Beech, Hazel and Ash all being utilized in this regard. Hodgkinson (2013) adds that a readily available source of water was also important in the location of individual sites, he saying: *"A lot of the early iron-working sites are found in stream valleys. This provided water to support all aspects of the operation, with the added bonus that it also facilitated prospecting along the banks of the streams."*

As detailed above in Section 3.1.2, the iron industry in the Weald had its origins in the LIA, and its success in that regard was a key factor in attracting Roman interest from the outset of the occupation. Data from sites such as Beauport Park (Brodribb *et al*, 1988, 232) show that from these comparatively modest beginnings the iron manufacturing industry expanded rapidly from that point. Hodgkinson (2008, 2) concurs with this chronology, saying: *"...iron making in the pre-Roman and Roman Weald should be regarded as a continuum that was unbroken, but intensified, by the Roman occupation..."*

Hodgkinson (2008, 32) says that there was considerable variation in the layout of Roman iron working sites. For example, the standard *chaîne opératoire* process for iron production at the smaller sites would typically find them based in a stream valley with the ore being dug from mines at the top of the valley slopes, then moving downhill to be roasted, then moving further downhill for smelting and forging before the waste was dumped into the valley bottom. As Hodgkinson (2013) adds regarding these smaller sites, *'everything moved downhill'* (see Figure 17). In contrast, the larger sites (especially in the eastern/ coastal region) would have been much more akin to Peacock's (1982, 8) manufactory-scale mode of production in the pottery industry. This would have been in terms of their large size, engagement with regional transport infrastructure and the level of industrial organisation evident (in the case of the latter, with a clear symbiosis between capital and labour).

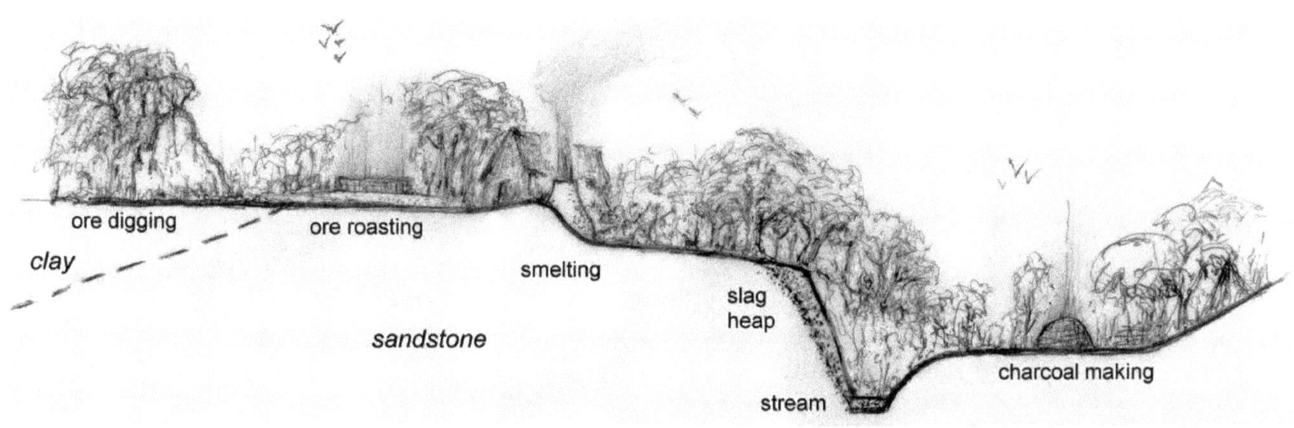

Figure 17: Cross-section of a typical smaller Romano-British bloomery site in the Weald. Reproduced from *The Wealden Iron Industry* (Jeremy Hodgkinson), original image by R. Houghton.

The siderite iron ore extraction operations would have been in the form of shallow quarries or bowl-shaped opencast pits, with the largest iron ore mining sites being located at Bardown and Beauport Park (the latter being the largest, Cleere and Crossley, 1995, 15). Hodgkinson (2008, 13) does emphasise however that iron ore would have been sporadically available across the whole of the Weald, and in many cases would have been mined very close to the iron working sites themselves, for example at Footlands Farm (Cleere and Crossley, 1995, 303). This is in contrast to the later occupation period iron manufacturing industry in the Forest of Dean where the centrally mined ore appears to have been shipped across the region using the River Severn as the main arterial routeway (J. R. Allen, 2010, 41).

In the Weald, once extracted the ore was then roasted to create ferrous oxides from the carbonate, this being easier to smelt. Furnaces came in a variety of types and were generally larger than any of their chronological successors until the later Middle Ages at the earliest (Dark and Dark, 1997), with Hodgkinson (2008, 2) saying: *"... both tapping and non-tapping furnaces were used, (with) both domed and shaft furnaces (being) found in the region during the same period."*

The Roman iron-workers would have used the 'Direct Process' when producing their iron, with the iron produced in the furnace being available for forging immediately (see Figure 18). The iron was obtained from these furnaces by creating temperatures of around 1,100 degrees centigrade, the slag then being removed and discarded and thus providing the principal evidence today of the occupation-period iron industry (along with other waste products including charcoal refuse, ore refuse and furnace debris). In his most recent calculations, Hodgkinson in work in preparation (pers. comm. 5 December 2015) has used this slag to determine the total amount of iron produced in the Weald during the occupation. Working with Crew's (1998, 51) estimates for the occupation-period iron industry at Laxton in Northamptonshire, which suggested that the total iron produced was equivalent to 15% of the total slag volume/ weight (the two amounts reckoned to be equivalent), Hodgkinson now indicates that the slag volumes in his 1999 and 2008 papers should be modified down by as much as 50% and that a more realistic figure (including an amount for a percentage of undated sites) may be nearer 75-100,000 tonnes of slag and waste in the Weald, indicating a total of 10-15,000 tonnes of iron being produced.

In terms of the output of each site, a large operation such as that at Bardown would have used eight furnaces to produce up to 50 tonnes of iron annually from 290 tonnes of iron ore. Cleere and Crossley (1995, 78) explain that to maintain this intense output some 15ha of woodland would have been needed to provide the annually required charcoal, with Hodgkinson (2013) believing coppicing would have been mandatory to maintain this level of woodland exploitation. Each site was extensively exploited for the locally available raw materials, and once these (particularly the iron ore) were exhausted satellite sites would have been established to make use of the existing supporting infrastructure, with the workers still likely to have been based at the original site (Cleere and Crossley, 1995, 72). A good example can be found at Bardown where the main site fell out of use by the end of the 2nd century AD but where satellites (including the location of the High Weald coin hoard, see below) continued well into the 3rd century AD.

Cleere and Crossley (1995, 81) believe that the iron produced by this intensive industry would have been used for four categories of goods, these being:

- Tools and implements.
- Weapons.
- Construction ironwork (nails, carpenters dogs, clamps and similar).
- Miscellaneous (horseshoes, boat fittings, barrel hoops and similar).

Additionally, the iron slag which was a by-product of the manufacturing process was also used in compacted form as metalling for Roman roadways in the region. As an example of the scale of this use, Hodgkinson (1999, 68) says: *"Of the Roman road that runs north from Lewes,*

Figure 18: Artist's impression of Roman bloomery furnaces in Little Furnace Wood, Mayfield, the Weald. Reproduced from *The Wealden Iron Industry* (Jeremy Hodgkinson), original image by R. Houghton.

Sussex, across Ashdown Forest (and) towards Edenbridge, Kent, as much as 30km may be surfaced with slag".

Post-LIA iron production seems to have begun early in many of the eastern/ coastal sites such as Beauport Park (Brodribb *et al*, 1988, 232), and also at a few of the more westerly inland sites such as Great Cansiron (Tebbutt, 1971, 11), peaking across the Weald in the middle of the 2nd century AD by which time the above mentioned 114 sites were or had been operational as iron working or iron-industry supporting locations. Changes in this prime example of the exploitation of natural resources in the region become evident however from the beginning the 3rd century AD, with Jones and Mattingly (1990, 193) saying:

> *"The major workings at Bardown and Crowhurst Park appear to have run down while satellite sites developed around them. In the mid-*3rd *century AD mining seems to have ceased altogether in the Bardown/ Holbeanwood complex and in the Battle area around Beauport Park. Therefore, the major production period for this industry...probably had a life span of approximately two centuries or a little more."*

More recent analysis of each of these sites as detailed above in Section 3.1.4 shows this still to be an accurate assessment, for example by Hodgkinson regarding Bardown (2012, 1). Cleere (1977, 18) also emphasised the abruptness of the end of iron production at these key sites in the middle of the 3rd century AD, while Cunliffe (1988b, 86) added that the maritime infrastructure around Romney Marsh built to support the iron industry also disappeared in the 3rd century AD. More recently Booth (2001, 3) has similarly shown that activity of all kinds, including iron production, declined dramatically at the roadside settlement of Westhawk Farm in the same timeframe. Meanwhile Stuart-Hutcheson (2012) says that the end date for the High Weald coin hoard found in 2006 near Bardown also correlates with the decline of iron production there at this time. This hoard is particularly important because of:

- The unique location of its finding in the central High Weald. The limited number of hoards found in the Weald until this point, for example that found in Hastings in 1989 (Dance, 1995, 3), are from the periphery.
- Its size, with 2,891 radiates dating from AD 215 to AD 268 and featuring all Emperors from Caracalla to the Gallic Emperor Postumus apart from Severus Alexander and Maximinus. By way of comparison, the Hastings hoard featured 59 silver *denarii* and 92 bronze coins (Dance, 1995, 3).
- The comparatively high number of rare coins, for example a radiate of Gordian III's wife Sabina Tranquillina and coins documenting the Secular Games of Phillip 1. Stuart-Hutcheson (2012) interprets this as representing high status.

- The range of mints represented, including Rome, Antioch, Milan and Lyons, indicating a mercantile origin.

Meanwhile Lyne (1994, 545) also highlights an economic change in the nature of pottery manufacturing in the region in the mid-3rd century AD, where he says a change from wheel-worked high quality pottery to handmade ware producing industries is directly associated with the decline of the iron industry. He suggests that the skilled pottery industry workers may have actually dispersed to work elsewhere in occupied Britain.

Some iron working does appear to have continued in the Weald after the middle of the 3rd century, mainly at a local level at the westerly sites which were less associated with the *Classis Britannica* (see discussion in Chapter 6). Cleere and Crossley (1995, 81) estimate that annual production figures for iron across the entirety of the Weald between AD 350 and AD 400 would have been 50 tonnes, compared to a peak of 750 tonnes between AD 150 and AD 250 when the changes described above become visible. They say that this later, more localised phase of iron-production continued into the 4th century AD by which time only Oldlands and Broadfields in the central region and Footlands Farm in the eastern/ coastal region were still active (and then at much reduced levels). Hodgkinson (2008, 34) says that any evidence at all of iron manufacturing, even at this limited level, disappears totally with the end of the occupation. For a long time afterwards there is no evidence of iron working at all in the Weald, and when it does reappear in a 9th century AD Saxon context at Millbrook in Ashdown Forest the technology in use is actually more primitive than that used in the LIA prior to the Roman occupation (Hodgkinson, 2008, 35).

Cleere and Crossley (1995, 81) make the case that one of the reasons for this evident decline of the occupation Wealden iron industry in the 3rd century AD was the silting up of the region's rivers which had provided the crucial access to the coast, especially with regard to the major sites in the eastern/ coastal region. Given the location of the latter sites near the coast, they were also highly exposed to attack by Germanic pirates (Harrington and Welch, 2014, 109), with the appearance of four Saxon Shore forts between Dover and Portchester being testament to the level of this threat (though noting Cotterill's contention is this regard, 1993, 236). Rudling (2013) suggests that over-exploitation of the Wealden forests for timber would also have been a factor in the decline of the industry given the intensity of iron manufacturing at its height (particularly in the eastern/ coastal region), while another factor was the decline in the demand for iron using the east coast maritime route, for example in London or by the military in the north. With regard to the latter, it certainly coincides with the post-Severan campaigns period in Scotland after which Southern (2013, 251) argues an unusual period of comparative peace followed which lasted for around four decades. She says in this regard that the 'slash and burn' policy of these campaigns led to severe de-population in the region which took several generations of peaceful co-existence to overcome. Meanwhile, the additional synergy between the decline of the iron industry and the disappearance of the *Classis Britannica* (particularly in the eastern/ coastal region) is discussed in full in the discussion in Chapter 5.

3.5 Who Were the Workers in the State-Run Areas of the Wealden Iron Industry?

While there is extensive data available to enable the consideration of the occupation-period Wealden iron-industry, little attention has been paid to who the workers actually were in the region. I do that here. As detailed in Section 3.1.2, a number of leading commentators (for example Brodribb, 1979, 141, Cleer and Crossley, 1995, 64, and Hodgkinson, 2009, 34) believe that the state had a significant presence in the eastern/ coastal region, with the *Classis Britannica* often being cited as the actor in that regard. This view provides context for the below, though is actually discussed in detail in Chapter 5.

At the top of the tree in the Wealden iron-industry (particularly the eastern/ coastal region) would have been the *procurator metallorum,* a common figure across the Empire at any mining and quarrying site (especially those state-run) and a senior figure within the Imperial economy. Often those recorded have the location of their area of responsibility attached to their official title, whether for an entire province, region or just for one site. An example is provided by one M. Ulpius Eutyches who is recorded in the early 2nd Century AD as the *procurator metallic Albo-C(rarensis)*, a mining district in north western Spain (Hirt, 2010, 120). The *procurator metallorum* had wide responsibilities which covered not only the successful operation of the *metalla* for which they were responsible, but also all ancillary activities. As de la Bédoyère (1992, 100) explains, this would have included transport and also the approval of the concessionaries who provided all related services, these latter ranging from the provision of food to supplying clothing and footwear.

One trend which does appear to be common with these senior officials relates to their societal origins. The more senior *procurator metallorum* (operating at either provincial level or looking after a region or a major individual site) were equestrians, while those managing the more minor sites tended to be freedmen (Hirt, 2010, 165). This distinction was also reflected in the size and seniority of their management teams, with those of the more senior being made up of junior *procuratores*, other equestrians or freedmen, and any military personnel seconded by the Governor (see Sections 2.2.2 and 2.2.4 for discussion regarding the various organisational options available to manage each specific *metalla* enterprise).

An interesting reflection here is where these *procurator metallorum* and their staff would have lived when not carrying out their daily duties. Certainly not close to their place of work given the toxic working environment, with Darvill and McWhirr (1982, 137) explaining that the *metalla*: *"...required a considerable amount of land, produced noxious smells and the movement of their products or even raw materials would have caused congestion on roads (or rivers) at peak times of production."*

This raises a problematic issue in the Weald, given the comparative lack of elite housing in the region (at least as found to date) to accommodate the procuratorial staff, excepting around the fringes as detailed in Sections 2.3.4 and 3.2. One intriguing possibility is that they actually resided, at least part of the time, in the grand villas of the upper Medway Valley. The Roman road running from Rochester through modern Maidstone to the major iron-manufacturing site at Beauport Park in the eastern/ coastal region of the Weald is well known as detailed above in Section 3.3. However, new data is now beginning to emerge regarding another Roman road which runs from the Dean Street ragstone quarry above the Medway Valley to the East Farleigh villa site/ Barming Roman ford on the River Medway (see Section 5.1.4 for full details). Crossing this ford one is then only a short distance from the villas of the north bank of the river, for example at Teston. Research is now being carried out to see if a link can be made between the eastern end of this road at the Dean Street quarry site and the close-by Wealden road which is only a short distance away. If such a link were found then this road turns into a spur of the major north-south route and gives a direct, physical link between the *metalla* of the Weald and the Medway Valley. That being the case a reasonable argument can be made that the *procutaror metallum* in Kent was one individual covering the *metalla* of both the Weald (siderite mining and iron manufacturing) and Medway Valley (ragstone quarrying), having as his residence a grand Medway Valley villa, rather than there being two such individuals, each covering their own respective area. A single individual would clearly speak to the importance of the region in an industrial context, particularly if the post was created early in the occupation which may account for the rapid growth of both *metalla*. Such an individual could also of course have resided in London (at least part of the time), travelling as required down the Thames Estuary and up the River Medway to reach the ragstone quarrying industry, then using the Rochester - Wealden road if he needed to access the iron manufacturing industry in the eastern/ coastal Weald.

Looking below the level of the *procurator metallorum* and his staff who ran the *metalla* at provincial, regional and individual mine/ quarry level, there were a range of options available to directly manage the large-scale *metalla* operations (though noting all would report upwards through the *procurator metallorum*). These ranged from the state carrying out the work using military resources, through to all the work being outsourced to civilian contractors (again, for the specific options see Sections 2.2.2 and 2.2.4). In the case of the former the discussion in Chapter 5 considers whether this was the case with the *Classis Britannica* for the eastern/ coastal area of the Wealden iron industry (and indeed elsewhere in the region, taking into consideration the *vilicus* detailed in Section 3.1.4) and whether this specific example was actually an Imperial Estate. In the case of the latter a good example would be the south western lead mining industry detailed above in Section 2.6.3, or the Forest of Dean iron industry where Holbrook (2006, 124) argues that the rights to extract and work the minerals were quickly granted by the state to local entrepreneurs and where Jackson (2012, 169) says that data from the key excavations at the small town of *Ariconium* show little sign of a military presence.

Going down the employment chain, looking at who actually carried out the physical siderite mining in the shallow quarries and bowl-shaped opencast pits, there are four possibilities.

- Military personnel if the sites were run directly by the state (as part of the Imperial economy, though clearly also contributing to the provincial economy, see Sections 2.2.2 and 2.2.3). Hirt (2010, 197) argues against this however, using data from mines and quarries across the Empire to show that while the soldiers and sailors were expected to physically participate in stone quarrying and wood cutting for the construction of military and public buildings, they may have been spared mining for iron ore. He says: *"Besides the hazards of working in a subterranean environment, the extent and intensity of labour involved (in mining) may have excluded soldiers as a constant supply of work force."*
- Semi-autonomous individuals or groups of freedmen (Harrington and Welch, 2014, 110).
- Indentured workers from the local population working under imposed munera (required duties), with Hirt (2010, 227) arguing that most inhabitants of a district hosting a mining enterprise could be summoned by the procurator metallorum for such work. Mattingly (2011, 172) details the analogous Roman gold-mining experience in Spain, saying: *"Studies of the settlement pattern around the northern Spanish mines suggest a significant level of reorganisation of local communities to support the mining activity...backed up by the discovery of an edict of Augustus from El Bierzo."*
- Forced labour, for examples criminals condemned to the metalla and slaves (Pearson, 2006, 64).

We should not forget here the grim life experiences of these faceless and less fortunate workers carrying out the 'coal face' mining activity. Brown (2012, 8) talks of nine tenths of the population across the Empire living miserable lives, and those carrying out the manual labour in the siderite mines of the Weald would have led a particularly unfortunate existence, especially if indentured workers, forced labour or slaves. For the criminals and slaves in particular, the experience of working in the mines would have been particularly traumatic, with Mattingly (2011, 174) pointing to evidence of violence to the bodies of miners in the archaeological and historical record across the Empire.

Moving from the miners to those actually manufacturing the iron from the ore mined above, this was clearly highly skilled work and the use of troops and sailors cannot be ruled out here. Cleere and Crossley (1995, 75) argue that in the larger establishments it would have been a mix of naval personnel and Imperially-employed civilian craftsmen (see discussion in Chapter 5). The origins of these civilian iron workers remains problematic, with Cleere and Crossley (1995, 75) believing the original LIA iron workers, who pre-dated the Romans at sites such as Garden Hill, Crowhurst Park and Footlands Farm, originated in *Gallia Belgica*. They then argue that as the state took over the industry, expertise from iron-working areas elsewhere in the Empire (for example *Noricum*) would have been imported into the region to facilitate the rapid growth and introduction of new technologies.

At its height during the occupation the total population in the Weald is estimated by Cleere and Crossley (1995, 75) to have been around 3,600 people, of which 600 would have been working directly in the iron industry (both mining and manufacturing), 600 in the supporting administrative and transport industries and the remainder either the suppliers of services or families. As detailed in Section 3.1.2, Hodgkinson (2008, 32) believes that many of the individuals working at the smaller sites in the central Weald were actually seasonally employed and may have been local farmers who carried out the activity out of season.

In terms of accommodation, while the *procurator metallorum* and his staff would have found fine living on the periphery of the Weald or in the Medway Valley, the majority of the extensive labouring workforce in the Weald would have been housed near to their places of work (Hirt, 2010, 46) and would have had to live with the harsh conditions there on a daily basis, especially the unskilled workers. In this regard, and based on data from Bardown, Cleere and Crossley (1995, 303) argue that the majority of this wider workforce would have lived in timber framed barrack-type buildings.

3.6 The Tile and Brick Industry in the Weald

In addition to occupation-period iron production, the Weald is also heavily associated with tile and brick manufacturing. This industry, which existed side by side with its iron producing counterpart, is perhaps best known for the tile and brick found across the region featuring the stamp of the *Classis Britannica,* this association with the iron industry and regional fleet being the reason it is considered in such detail here. The first such stamped tile was found in 1778 by the Revd. John Lyon in Dover, and from such modest beginnings subsequent archaeological investigation has cast light on what we now know was a major influence on the success of the economy in the region.

Mills (2013, 453) says there is no evidence of tile and brick manufacture in quantity in Britain before the occupation. Its appearance is one of the first signs of the arrival of *Romanitas* here, and the Weald was one of the first areas this industry emerged after the AD 43 Claudian invasion, perhaps as early as AD 50 (though at first small in scale). This clearly illustrates a similar degree of foreknowledge regarding the available extractive materials as is evidenced by the Wealden iron industry and the Medway Valley ragstone quarrying industry. Mills (2013, 461) says that the military and settling veterans were the principal vector for the introduction of tile use in Britain, with Hirt (2010, 197) arguing that the soldiers and sailors would have physically participated in the manufacture of tiles in the same way they would have carried out stone quarrying and wood cutting.

The first municipal or official tile stamps in Britain appear in AD 90 (in a legionary context), with Betts (1987, 28) arguing that this late appearance of official stamping shows that the very early tile manufacturing industry would have been local in nature rather than extra-regional. This is illustrated by Table 3.2 which sets out the location of kilns used in the early phases of the urbanisation of London, with the Weald not featuring at all (tiles from there beginning to appear in the provincial capital after the beginning of the 2nd century AD).

The nature of the Wealden tile industry paralleled that of the iron industry, with the data indicating similarly local operations in the central Weald and larger *Classis Britannica* associated state-run brickyards in the eastern/coastal region. The tile industry seems to have matured somewhat later though, peaking in the later 2nd century

Table 3.2: Locations of kilns producing tiles used in London early in the occupation

AD 50-80	AD 70-100
Ash	Canterbury
Canterbury	Deerton Street
Deerton Street	Turnham
Turnham	Maidstone
Maidstone	Eccles
Eccles	Rochester
Snodland	Ebbsfleet
Ebbsfleet	London
Darenth	
Orpington	
Beddington	
London	
Romford	
Chelmsford	
Heybridge	
Great Holst Farm, Boreham	
Colchester	

Betts, 1987, 28.

AD. By the time it did however, the eastern/ coastally-produced *Classis Britannica* stamped-tiles in particular were ubiquitous around the entire east Kentish coast (and indeed further afield, see Table 3.2), with buildings featuring this mark often being associated with a state-related function (and providing one of the principal means of identifying such sites as having a *Classis Britannica* provenance, see discussion in Chapter 5). In terms of utility, Brodribb (1979, 141) argued that the official stamp of the regional navy was used on the state-produced tiles for prestige reasons and to ensure their quality. In that regard it is noteworthy that the stamps always appear on the upper sides of *tegulae, imbrices* and floor tiles (the stamp also appearing on bonding, hypocaust *pilae* and box-flue tiles) rather than their underside, and would thus have been externally visible.

The process of tile manufacture was standardized across the Empire, though clearly there would have been regional variations. McWhirr and Viner (1978, 360) summarised the Roman tile manufacturing process as detailed below:

- Clay excavated, usually in the autumn.
- Clay allowed to weather over the winter, being broken down by frost and rain.
- Clay prepared for manufacture, with aggregate sometimes being added and then the finished product being covered until needed.
- Tiles made using a wooden frame, mold or former, when any modifications such as the flange on *tegulae* would be added.
- Tiles left to harden, sometimes being stamped.
- Tiles fired or burnt in a kiln or clamp and then stored. The iron oxides in the clay, and the conditions of firing, determined their colour.
- Tiles transported to buyer and used.

Two types of tile with *Classis Britannica* stamps were identified by Peacock (1977, 236) in his extensive survey which is still the benchmark to this day, this based on their material composition and which he styled Fabric 1 and Fabric 2. The Fabric 1 type is a uniform buff colour with scattered quartz sand-grains and is identical to unstamped tiles found extensively around Boulogne. Peacock believed they were manufactured in a brickyard in the vicinity of the *Classis Britannica* headquarters there, from material quarried locally, and were used regionally with very limited export taking place. Until recently the only tiles of this fabric found elsewhere (over 100 were found around Boulogne and a small, unknown number nearby at Desvres, Brodribb, 1970, 1) were an example from Dover and another single tile from the central-Wealden port and iron-working site at Kitchenham Farm. However, more recent research by HAARG (see Section 3.1.4 above) has dramatically increased the number found at the latter location, with 29 out of 31 additional pieces of *Classis Britannica* tile of this origin recently coming to light (Cornwell and Cornwell, 2008, 10, 2008, 1, and 2010, 16). This is in itself a strong indicator of trade taking place between Kitchenham Farm and the continent, and a good indicator of the extant levels of market integration.

Figure 19: Fabric 2 *Classis Britannica* stamped tile from the Winchester Palace site in Southwark. Ian Betts.

The Fabric 2 tiles are very different to those of the Fabric 1 type and represent those manufactured by the *Classis Britannica* in the eastern/ coastal Weald. Given their importance to this research they warrant a full description, with Peacock (1977, 237) saying:

> *"Fabric 2 is of a distinctive reddish-pink colour relieved to a degree by streaks, lenses and swirls of creamy white clay. The most prominent and abundant inclusions are of black or red-brown ore, usually as near-spherical particles about 1mm across: but irregular rounded fragments up to 3mm or more across are usually present. Sub-angular fragments of white, finely laminated siltstone (up to 10mm) are another feature and a few pieces are almost invariably present in a cross-section 20cm long. In contrast to Fabric 1, quartz is not visible in the fractured cross-section, but sand- or grit-size fabrics are occasionally seen on or near the surface, where they result from dusting to prevent the clay sticking to the mold during manufacture."*

Peacock (1977, 237) believed that the material used for these tiles was Fairlight Clay which sits within the Ashdown Formation in the Hasting Beds. This material appears a very good match for that used in the Roman tiles and Peacock noted that, while such tiles were used across a wide variety of sites across the region, the area where the vast majority were found was in fact around Fairlight (located in the eastern/ coastal region). He specifically believed that the principal naval brickyards used were located near where the Fairlight Clay actually outcrops at Fairlight Head, or from deposits beneath the alluvium of Romney Marsh between Rye and Dungeness. In the first case he argued there is a good chance that the principal brickworks have been lost through coastal erosion, while in the second he believed the evidence would have been covered over by recent alluvium deposition in Romney Marsh.

Table 3.3: Distribution table, locations and totals of *Classis Britannica* stamps on tile of known provenance (in alphabetical order, all Fabric 2 unless specified, together with site history; no distances given as exact place of origin ill defined)

Location	Number of Stamps	Site History
Bardown – eastern/ coastal Weald	28	The first *Classis Britannica* tile was found here by Brother Stephen Pepperell in 1951 and identified by Margary (Brodribb, 1969, 109). All subsequent tile was found during the excavations of Cleere in the late 1960s.
Beauport Park – eastern/ coastal Weald	1,600	The vast quantity of stamped tile found here was discovered during the investigations of Brodribb and Cleere in the 1970s and 1980s, most in association with the well-recorded bath house (Brodribb et al, 1988, 268).
Bodiam – eastern/ coastal Weald	31	Found between 1959 and 1969 in excavations by the BDHS (Brodribb, 1969, 111).
Boulogne	100 plus (approximately two thirds Fabric 1)	Mostly discovered in the 19th century, many of these tiles have since been lost though a number remain in Boulogne Museum (Crowley and Betts, 1992, 218).
Desvres	?	*Classis Britannica* tiles were reported at this site inland from Boulogne in the antiquarian record but the provenance of their finding is unclear (Crowley and Betts, 1992, 218).
Dover	1,000 plus (one Fabric 1)	First tile found in 1778, with many since, particularly in the context of the excavations of the *Classis Britannica* fort and the 'Painted House' (Philp, 1989, 57, and 2014, 38).
Folkestone	21	Seven tiles featuring a *Classis Britannica* stamp were found during the initial excavations of S.E. Winbolt in 1923-24 (Brodribb, 1969, 109). Since that time others have been recovered from successive excavations and from spoil (much from the first excavation) at the foot of East Cliff, East Wear Bay.
Kitchenham Farm central Weald	31 (29 Fabric 1, most in pristine condition)	HAARG's recent investigations here have located 31 *Classis Britannica* tiles, 29 of which are of the Fabric 1 type originating in Boulogne. These tiles are the first and only ones located to date in the Central Weald.
Little Farningham – eastern/ coastal Weald	53	Site located by the landowner in the 1950s and excavated by the Cranbrook and Sissinghurst Local History Society.
London	5	*Classis Britannica* stamped tiles have been found here at Nobel Street, Winchester Palace (Crowley and Betts, 1992, 218, see Figure 19), Hunt's House (Taylor-Wilson, 2002, 10) and Garlick Hill.
Lyminge	2	Two re-used tiles have been found in the context of the Saxon palace under excavation by Reading University (unpublished at the time of writing, original provenance unknown).
Lympne	22	All *Classis Britannica* tiles located here were found during the Charles Roach-Smith excavations of the Saxon Shore fort in 1850. Others may have been found subsequently but have been lost (Brodribb, 1969, 108).
Pevensey	3	These tiles were found during the investigations at the site of Sussex antiquarian L.F. Salzman in 1906-1907. Others were apparently found between 1907 and 1939 but were lost at the outbreak of the Second World War (Brodribb, 1969, 109).
Richborough	1	One tile found in 1932 (Atkinson, 1933, 10).
St Catherine's Point, Isle of Wight	1 (Fabric 1)	Lyne (2000, 10) speculates that this single example found in a late period enclosure ditch may indicate the previous presence of a signal station on the southernmost point of the Isle of Wight.

Brodribb, 1969, 185, 1970, 25, and 1980, 191, Cleere and Crossley, 1995, 65, and Crowley and Betts, 1992, 218, updated by Elliott, 2015.

While predominantly found in the eastern/ coastal region of the Weald, the use of Fabric 2 tile was widespread as is evident in Table 3.3 and Map 3.2, with Peacock believing that of the two regions manufacturing *Classis Britannica* tiles, that making the Fabric 2 tiles was by far the most important.

For completeness Rudling *et al* (1986, 227) also identified a number of early non-*Classis Britannica* tileries in the Weald, mostly in the central region, for example that at Great Cansiron which operated from the late 1st into the 2nd centuries AD. Here some of the tiles were stamped with a roller design which the authors speculate might indicate the presence of a group of itinerant tile manufacturers who had relocated here to support the growing local iron manufacturing industry. Black (2013, 41) says that while early in the occupation these civilian contractors were readily used by the state at military and industrial sites, by the beginning of the 2nd century AD those manufactured in the *Classis Britannica* brickyards in the eastern/ coastal

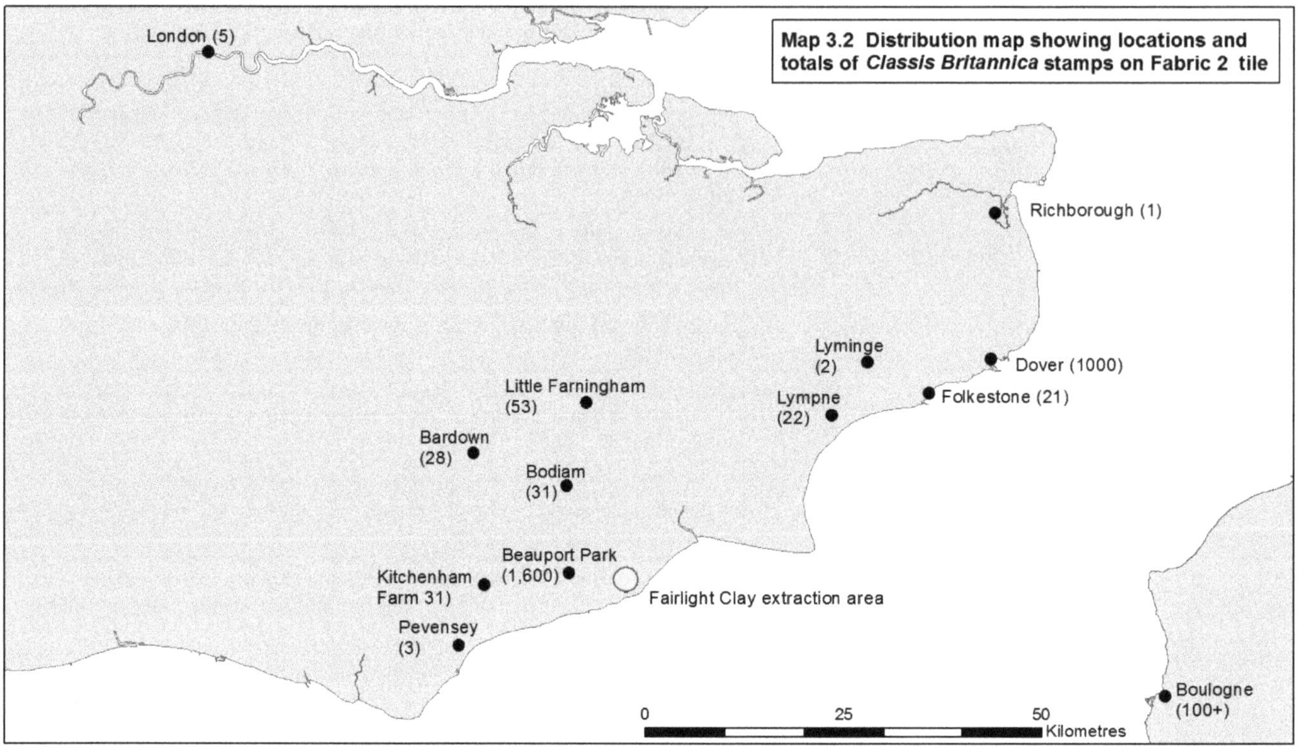

Map created by Elizabeth Blanning © Crown Copyright/database right 2016. An Ordnance Survey/EDINA supplied service.

region predominated, with the state tile industry reaching critical mass around this time. Peacock (1977, 245) believed that Wealden Fabric 2 brick and tile manufacture may have actually begun in the late 1st century AD to facilitate the building of regional state infrastructure such as the twin *pharos* lighthouses at Dover, these brickyards therefore operating alongside the early, centrally-located tile manufacturing sites for a short time.

In terms of the longevity of the industry, data from across all the sites where *Classis Britannica* tile has been located supports the theory that it continued to thrive while the iron industry in the region was still running at full capacity, then declining after the mid-3rd century AD. It was certainly still in operation at the beginning of the 3rd century AD when newly manufactured Wealden *Classis Britannica*-stamped box-flue tiles were being used to build the 'Painted House' *mansio* in Dover (Philp, 1989, 101).

4

Regional Analysis – Medway Valley

This chapter forms the core new research here, focusing on the occupation-period ragstone quarrying *metalla* of the upper Medway Valley in Kent (see Figure 20). This was an industrial-scale operation of a size not to be replicated until the modern era, supplying through the River Medway, Thames Estuary and South East coastline much of the building material used in Roman London and the region through to the mid-3rd century AD. The research presented here is specifically relevant to the debate about the state presence regarding this and other regional industries, in the context of the Imperial and provincial economies. The chapter begins with my detailing the key primary evidence sites in this region (including background on the River Medway given its central role with regard to this *metalla*), followed by an analysis of this upper Medway Valley ragstone quarrying industry, a further analysis of where the specific ragstone quarries were physically located (revealed here for the first time as part of this research), a reflection on who the workers were that serviced this *metalla*, before finally presenting a discussion on the transportation of the quarried ragstone (including, again for the first time, the recreation of a typical journey of a boatload of quarried stone from quarries identified here to Roman London).

4.1 Key Data: Details of Primary Evidence Sites

Here I once again set out the key data from all of the primary evidence sites, preceding the site-by-site analysis with a discussion on the nature of the evidence, a review of the origins of the exploitation of natural resources in the region and a comment on site selection, all to provide understanding for the reader in the ensuing site list.

4.1.1 Nature of the Evidence

Until the advent of this research, archaeological activity in the upper Medway Valley has been typified as a series of piecemeal discoveries and small to medium scale (though in some cases lengthy) excavations, with the detail and data from the research often not being academically presented. Crucially, no attempt was made until my MA 'Medway Formula' research to view the Roman-period valley holistically (see below).

The upper Medway Valley has long been associated with the quarrying of ragstone, with for example the antiquarian Lambarde (1576, 1826 edition, 174) saying that he had seen the county town of Maidstone called *Maegpanstane* (referencing strong stone) in an *'ancient Saxon book.'* Other antiquarians also referenced the provenance of the available extractive materials here, with Hasted (1797, 9, and 1800, 2) noting that the fertile soils of the valley covered 'quarry rock'. It is around this time that the first antiquarian references begin to appear regarding the occupation-period sites in the valley, for example at Barming where the then leading historian and biographer Rev. Mark Noble noted in 1797 the Roman buildings and cemeteries there (unpublished at the time, his private papers being first recorded by Roach-Smith, 1848, 183).

Figure 20: Beds of cream coloured ragstone and light brown hassock exposed in the upper Medway Valley. This is a typical bedding structure for the stone in the region. Simon Elliott.

Around the same time as Noble's early investigations Hasted (1797, 2) also highlighted the Roman provenance of Barming, citing finds there of Roman urns, armour and skeletons. Using Barming once again as an example, Smith (1839, 59) also noted in a guidebook to Maidstone that the walls and bath of a Roman villa were still visible there within living memory. In the same work (1839, 57) he also recorded the finding in 1838 of what we now know as the Roman villa in East Farleigh.

From this time such antiquarian investigations become more common in the Medway Valley, though in complete isolation and with the findings (when published at all) appearing in a diverse set of antiquarian publications. Roach-Smith (1844a, 117) and Wright (1854, 189) both record the finding of a Roman building in Snodland (again now understood to be a villa, and again with no context), Charles (1847, 86) details his investigations of the villa now known as The Mount in Maidstone which was first revealed in the 1843, and Grover (1873, 45) reports local hop-field owner Arthur Fremling discovering and excavating what we now know is the Roman villa at Teston in 1872.

Sporadic archaeological investigative activity continued into 20th century in the region, though the Medway Valley lacked a central figure to galvanise research in the manner of Straker in the Weald. Thus it was not until the 1960s that large scale, organised excavations began on some of the region's major Roman sites, for example that led by Detsicas with the Lower Medway Archaeological Research Group (LMARG) at Eccles from 1963 to 1976 (1967, 170). Similarly, The Mount villa site in Maidstone received significant attention in the early 1970s through a Department of the Environment-funded investigation by the Maidstone Area Archaeological Group (MAAG) led by A. Miles and D. Kelly (Miles, 1972, 217). The haphazard nature of regional investigations is evident at this site, with it next being revisited by CAT in 1994 when, for the first time, a full floor plan of the main occupation-period structure was created (Houliston, 1999, 71). Sporadic investigations of isolated sites has continued from that time to this day, most recently MAAG's ongoing investigations at the East Farleigh villa site (Daniels, 2015, 6) and the author and Wilkinson's (through KAFS) similarly ongoing excavations at the Teston villa site (Elliott, 2013, 40).

From the above narrative it is clear that, despite the Medway Valley featuring Maidstone and having a long history of settlement, agriculture and industry, there has been no organised approach to regional archaeological investigations. It was this fact that prompted the author to use the occupation-period valley as the focus of his MA dissertation research in 2009, in the context of the ragstone quarries which were known to have supplied much of the building stone for the South East during the occupation but which were at that time unlocated. The key elements of this research were:

- A discussion regarding maritime transport by river, canal and sea being the preferred means of transporting heavy goods in the pre-modern era.
- A discussion on the scale of the occupation-period ragstone quarrying industry in the upper Medway Valley above the tidal reach at Allington, where I advanced the hypothesis that this activity was industrial in scale, facilitating the urbanisation and later fortification of the South East of Britain (Marsden, 1994, 80). Further, I advanced an argument that the *metalla* along the upper Medway Valley was run by the *Classis Britannica* on behalf of the state, possibly in conjunction with the iron industry in the Weald.
- A discussion on change and continuity in this Medway Valley *metalla*, where I presented a hypothesis that the *Classis Britannica's* involvement lasted until the mid-3rd century AD (very much paralleling the experience in the Weald), after which the ragstone quarrying in the region became far more localised and smaller in scale.
- Finally, a discussion regarding the use of riverine hydraulic infrastructure in the upper Medway Valley during the occupation to facilitate access along the river for the quarrying industry, and the elites I argue ran it.

Overall I dubbed the broad findings of this research the 'Medway Formula' (this being described at length by the author in *Archaeologia Cantiana*, S. Elliott, 2014a, 251), its major outcomes threaded through this chapter where much new archaeological data is also considered given developments in regional archaeological research since that time, particularly in the context of my recent PhD thesis and also the recent research of others, for example Elizabeth Blanning (2014, 201).

As can be seen above, outside of the context of specific individuals, a wide variety of organisations have led the investigation of occupation-period sites in the Medway Valley. At an amateur level these have included KAFS, KAS, LMARG and MAAG, these being joined in the modern era by professional archaeological organisations such as Wessex Archaeology, Oxford Archaeology, Archaeology South East (ASE), CAT and KARU (assisted by strong volunteer input). The region has, in recent years, also benefited from wider community project investigations. Examples include the Lower Medway Valley component of the VCH's 'England's Past for Everyone' project (Hann, 2008), and the Randall Manor Community Archaeology dig led by Kent County Council (KCC) community archaeologist Andrew Mayfield (2015), this funded by a National Lottery Heritage Fund grant. Finally, interest in the occupation-period heritage of the Medway Valley has been assisted through the on-going activities of the region's plethora of local history societies.

Where recorded (and often they are not, at least in an academic sense), the findings of each generation's research of the occupation-period Medway Valley can be found in a variety of publications. From an early date these have included the *Journal of the British Archaeological Association*, the *Proceedings of the Society of Antiquaries of London*, the *Archaeological Journal* of the Royal Institute of Archaeology and of course *Archaeologia Cantiana*. To this were added as the decades progressed the KAS Newsletter, the CKA's *Kent Archaeological Review* and the KAFS Newsletter. Once again the key sites also feature prominently in the Kent HER and the 3rd Volume of the VCH History of Kent.

As with the Weald, there are specific features which define the archaeology of the Medway Valley. The first is the river itself which, through fluvial action, has frequently revealed Roman archaeology along its banks, for example at Snodland (Taylor, 124, 1932) and The Mount (Houliston, 1999, 71). Most recently the significant flooding event of Christmas 2013 led directly to the finding of 'the 'Medway Stones' (S. Elliott, 2014c, 11, detailed above in Sections 2.5 and 4.1.4). The second feature is that the region has seen more development in terms of settlement and particularly industry in the modern era than the Weald or the Folkestone region (significantly so in the case of the former). This has led to the finding of some of the key sites in the Medway Valley, for example the potential Mithraeum at Wouldham/ Burham (Jessup, 1956, 171). A final point to note is with regard to some of the antiquarian investigations of key sites in the region which should be approached with even more caution than usual given that where records of these early excavations do survive they are often in the form of commentaries by visitors rather than the reports of those actually carrying out the work.

In terms of the evidence used in the list of primary evidence sites, each one considered features some or all of the following: buildings, burials, pottery, coins, glass, material culture associated with industrial activity and other associated small finds. As with the Weald and the Folkestone region, dating throughout the period of investigation in the Medway Valley has largely relied on pottery and coins, and to a lesser extent glass.

4.1.2 Industrial Origins

There is no evidence of industrial activity in the Medway Valley prior to the Roman occupation other than for the manufacture of LIA pottery (for example flint-tempered ware from the Upchurch Marshes, Monaghan, 87, 26) and some pre-Roman coin moulds found during the 1961/ 1962 excavations on Rochester High Street (Chaplin, 1962, L, see Section 4.1.4). It is therefore clear that the industrial-scale ragstone quarrying and tile and brick industries which were to define this region so clearly during the first half of the occupation were innovations directly associated with the Roman conquest (S. Elliott, 2014a, 252).

Given the fertility of the valley and this early lack of industry, settlement during the LIA was almost entirely related to agriculture (S. Elliott, 2014a, 251), this also being evident in terms of continuity with some of the later elite Roman settlements which evolved from LIA predecessors. Examples include the villas at Eccles and East Malling where an LIA settlement context precedes later villas with an agricultural association (Blanning, 2013, 202). Other villa sites also developed in the context of pre-existing LIA settlement, though with an association with occupation-period industry rather than solely agriculture. Examples include the villa at East Farleigh with its association with ragstone quarrying (S. Elliott, 2014a, 252, see Sections 4.1.4 and 4.4) where recent material culture finds including an LIA silver minim coin in association with Iron Age ditches indicate pre-Roman activity. In a grander sense, the *oppida* which Howell (2014, 38) argues existed at Boughton Monchelsea also later found itself at the centre of Roman industrial activity, yet again a ragstone quarry (S. Elliott, 2014a). Blanning (2013, 202) does highlight however that this continuity of settlement (whether in the context of agriculture or industry) from the LIA to the occupation at key sites is not visible everywhere in the Medway Valley, saying it is not the case for example with the villa sites at Snodland and the Mount in Maidstone.

4.1.3 Site Selection

I argue above that the Medway Valley was home to a wide variety of industrial activity during the occupation, dominated by ragstone quarrying in the upper Valley. I therefore include in Section 4.1.4 all of the key occupation-period sites in this region (based on the available archaeological, historical and analogous data, all detailed site by site below) given each would have had some link or association with the local extractive industries exploiting natural resources. The only exceptions are the actual ragstone quarries themselves which, given their importance to the wider research (they being identified here for the first time), are considered separately and individually in Section 4.3.

Before beginning the review of primary evidence sites, special reference should also be made to the two principal towns in the Medway Valley today, namely Rochester and Maidstone. In the case of the former, much is known about the Roman small town of *Durobrivae* (translating as fort by the bridge), located at the point where Watling Street crossed the River Medway and the location of at least two bridges during the occupation (see Section 4.1.4). The junior Kentish partner to Canterbury in the Roman period, Mattingly (2006, 287) believes Rochester was also a subordinate centre of regional Government, administering a western Kentish *pagus* (the smallest administrative district of a province).

Maidstone however is a complete enigma, the current county town of Kent being the subject of much speculation concerning its character during the occupation. Various hypotheses range from it being a Medway Valley small town to a ribbon development of villa settlements and associated cemeteries. Certainly its location, just above the tidal reach at Allington where the Rochester - Wealden Roman road meets the River Medway, lends itself to a classic transport node small town interpretation (Hingley, 1989, 24) associated with a riverine port operating as a break of bulk point for the ragstone quarrying industry in the upper valley and the iron industry in the Weald (in the latter case using the Wealden road). The proximity of the town to the ragstone quarries also adds credence to a small town interpretation, de la Bédoyère (1992, 100) for example saying that many small towns owed their existence to the presence of state-run local extractive industries due to the need for mercantile activity to support the industrial operations, and also to provide accommodation for all levels of the workforce. Everitt (1986, 86) stated just such a case for Maidstone, saying he believed that the name for the town's principal thoroughfare, Week Street, is derived from the Roman *vicus*. This is a settlement type usually associated in the British Romanist tradition with a military site, though in France and the Low Countries more broadly with a small town. Detailed analysis of Roman Maidstone has been hindered however by urban growth in the town centre, particularly along the path of the Roman road (modern Sandling Road, then Week Street and on to Gabriel's Hill). Nevertheless Everitt (1986, 86) was still able to detail over 40 major archaeological sites on the 2.5" Ordnance Survey (OS) map of the town at the time of his research, more than any other site in Kent apart from Milton Regis and Ospringe, he believing that a significant urban centre lay beneath the modern town.

Maidstone is not however recorded in any occupation period itinerary, even though many antiquarians attempted to associate it with the *Vagniacae* in the Antonine itinerary (Robertson, 1883, 68) which is now known to be the Roman name for the religious small town at Springhead on Watling Street (Philp and Chenery, 1997, 3). In fact the first historical reference to Maidstone is a passing mention in the *Textus Roffensis* from around AD 975, while another interpretation of Everitt's linkage of Week Street with *vicus* might actually be with the Saxon *Wyke,* meaning village or hamlet (confusingly itself a derivative of *vicus*) indicating a Saxon origin. Certainly the modern town itself is a medieval founding (Cooper, 2008, 2). Observational evidence is also relevant in this debate as, to my mind, if Maidstone were a small town, then given the nature of archaeological activity in the area over the past two centuries one would have expected more non-villa related stone-founded structures (or even evidence of substantial wooden buildings) to have been located. To that end, even taking into account the hit-and-miss nature of archaeology within the modern built environment, the fact they have not is particularly telling. Recent excavations in Church Street, Maidstone, by CAT supports this view, with Weekes (2014, 149) saying: *"What we might call a small town, comparable with the Westhawk Farm settlement at Ashford, is certainly a step too far on present evidence."*

This is certainly the case when one views the dispersal of Roman sites in the area, this showing a sequence of villa estates at (going upriver) Allington, Little Buckland Farm, The Mount and Florence Road/ Bower Lane, and additionally along the Wealden road at Barton Road. The main cemeteries/ burials are clearly associated with these villas, particularly the large one at The Mount, which may be testament to the size of this estate. Therefore, given that there is no holistic data to support the small town interpretation other than a limited amount of analogy, I believe it is reasonable to determine that Maidstone was actually a series of medium frequency occasional villas and small-scale settlements strung out with an association with a major waterway and road, not a small town.

Finally here, before moving onto the site list itself, I detail the River Medway given its central importance to this chapter. The river is 113km (70 miles) long, rising in Ashdown Forest in West Sussex and flowing through East Sussex before transiting for most of its length through Kent before entering the Thames Estuary near Sheerness. In the antiquarian record (Grover, 1873, 10), and indeed more recently (Worcester, 2013), it is referenced as being called the Madus during the occupation and the Vaga prior to that, but there is no firm data available to support this. The current name is Saxon, referencing it as the 'middle way' given it bisects the county of Kent for much of its length.

The river was originally a tributary of the Thames when the latter flowed along its original course to empty into the North Sea off the East Anglia coast, with the Medway's current course being determined by the Anglian Stage Pleistocene glaciation (478,000 years to 424,000 years before present) which pushed the Thames south to its modern position (Pettitt and White, 2012, 109). There is no sign of any great autogenic processes having taken place in the period between the Roman occupation and the modern era, with the river following broadly along the same course. The most visible changes in the intervening period would be the flooding of much of the original estuary in the lower Medway Valley and creation of the current salt-marshes in the region, with Evans (1953, 105) saying: *"The lower Medway Valley presents an impressive example of a typical drowned estuary similar to many associated with the North Sea basin."*

The Medway has tributaries rising in Ashdown Forest, the Weald and the North Downs, with major examples including the Eden, Bourne, Busty, Teise, Beult, Loose and Len. The river has the largest catchment area in southern England at 2,409km^2 (930 miles^2) and flows in a west-east direction before turning north at the confluence of the River Beult and then breaking through the North Downs at the Medway Gap. Its use for navigation is determined by the tidal reach and today the river is tidal through to Allington Lock and Sluice except on very strong spring tides and then only for a few minutes when it can almost reach Maidstone. Above the tidal reach, without the current hydraulic riverine infrastructure in the form of locks and weirs, the Medway would effectively be a stream rather than the free-flowing river experienced today.

Geologist Kaye has argued recently in *Archaeologia Cantiana* (2015b, 232) that the tidal reach in the Roman period would have been even further downriver at Snodland. He has based this on an analysis of alluvium deposits on the riverbed, with factors influencing the subsequent shift to Allington being a balance between rising sea levels (Hall and Merrifield, 1986, 18 and Devoy, 1990, 17) and water entering the river from modern water treatment works along its length on the one hand, set against silting (especially in the Medway Estuary, see above) and water being removed for industrial and agricultural use on the other. The Medway enters the historical record in the 13th century in the Magna Carta, one of the latter's stipulations being the removal of weirs (*kidelli*) in the river (Ellis Jones, 2012, 29). Clearly little progress was made at the time, and in the 16th century it came under the control of the Commission for Sewers, initially from Lombards Wall to Sheerness and then from Sheerness to Penshurst Bridge. A major driving force in this regard was a desire to open up riverine access for the iron foundries along the upper Medway in the Weald to transport iron-cast artillery ordnance to London (Straker, 1931, 189). The activities of this Commission also appears to have been largely unsuccessful (excepting Cole's referencing the removal of weirs and stone shelves at East Farleigh, Barming and Teston, 1630, 134) as the river next appears as the subject of the 1665 '*Act for Making the River of Medway Navigable in the Counties of Kent and Sussex*'. Under this Act, Commissioners were again appointed to manage the river, though once more comparatively little activity appears to have taken place. However, in 1739 a new *Act 13 King George II chapter 26* facilitated a survey of the river and in 1740 the Medway Navigation Company was established with the aim of making the river navigable from Maidstone to Forest Row (and initiating the construction of the locks and weirs visible in the river today). Control of the lower, tidal reaches of the river remained with the Commissioners of Sewers. In 1802 however a further piece of legislation created an additional company known as the Lower Medway Navigation Company which finally replaced the Commissioners, and this dual arrangement continued throughout the 19th century, being extended by the new *Upper Medway Navigation Act 1911*. Finally, with the *Land Drainage Act 1930*, the responsibilities of the Upper Medway Company came under the control of the River Medway Catchment Board from the 1st April 1934.

The major crossing of the Medway is at Rochester, which currently features the town bridge, railway bridge and to the immediate south of these the east and west bridge crossings of the M2 and the high-speed rail link bridge. Crossings further upriver include the bridges at Maidstone, Tovil (foot), East Farleigh, Barming (foot), Teston, Yalding and Tonbridge, while substantial fords are detailed in the recent historical record at New Hythe, Aylesford and Tovil.

During the occupation, the major bridge crossing was at Rochester where Watling Street crossed the Medway, while ancient ford crossings are recorded in local tradition at Lower Halling, Aylesford, Maidstone (detectable at Lock Meadow where the river depth decreases dramatically for a short stretch), Tovil (similarly detectable today), East Farleigh, Barnjet, Barming and Teston (Brooks, 1994, 3). As I detail in Section 4.1.4, an additional Roman bridge may have been located at Tovil.

Moving back to the list of primary evidence sites, these are now detailed below heading upriver from Rochester.

4.1.4 Site List

Rochester Roman Bridge (at TQ 7416 6892)

Date: 1st century AD through to the early Medieval period (based on investigation of the exposed bridge pier in the 19th century, and pottery and coin data from Rochester).

Type: Riverine infrastructure.

Site History and Economic Evidence: Though long assumed to have existed by antiquarians given it was the Watling Street crossing point of the River Medway, the first hard evidence of the existence of an occupation-period stone built bridge here came to light in 1851 during the construction of the current cast-iron bridge in Rochester (Brooks, 3, 1994). When work began on the western Strood pier the engineers revealed almost immediately the ragstone remains of one of the Roman bridge piers (at TQ 7404 6895). Of this, the only bridge pier found to align with the 'modern' bridge, Hughes (1851, 365) detailed:

> *"...a mass of Kentish ragstone, of the nature of rubble without mortar, is found to a depth varying from 4m to 8m below the present bed of the river. Pieces of timber of considerable dimensions, and which had been used as piles, or framing (for a cofferdam), occurred in this bed of rubble stone, penetrating a foot or two into the gravel, which proved to be 2m to 2.5m thick. This timber was oak, elm and beech – all except that last perfectly sound and tough...some fragments of iron proved that the piles had been shod with that material."*

Brooks (1994, 3) argues that this structure was actually the second occupation-bridge at Rochester, the first being wood built. There is no hard data to support this however, though such a pattern of an initial wooden bridge being replaced by a later stone built structure was common elsewhere in the Empire (Brooks, 1994, 4). What we do know is that the stone bridge was indeed substantial, being 183m in length which is shorter than the London crossing of the Thames but longer than that of the Ouse at York. The then Bridge Engineer John J. Jobson said in 1921 (140) that he had seen documents (which he does not detail) which indicate that this stone-built structure lasted until AD 960 when it was pulled down, having ultimately become unsafe. Jobson (1921, 141) then cites other documents he had seen (again without giving detail) dating to AD 1115 which detailed that the Roman stone built bridge featured nine stone piers, two abutments and two openings (with draw-bridges) for the transit of high masted vessels. He argued (1921, 143) that this matched the bridge's later, Medieval replacement which he believes was simply a new wooden road surface built on top of the original Roman bridge piers.

Of relevance to this work, Yates and Gibson are clear in their belief that a bridge of the sophistication of the Roman stone-built structure would have been built and maintained by the state (1994, 5).

Rochester Town (centred at TQ 7427 6863)

Date: From at least 1st century BC through to the modern era (based on pottery and coin data).

Type: Small Town.

Site History and Economic Evidence: Roman Rochester has been the subject of numerous archaeological investigations, initially by antiquarians such as Poste (1859, 65) and Payne (1895a, 2), though it was not until 1960s that a full picture of the occupation-period small town began to emerge following investigations by Chaplin (1962, L). Further investigations by LMARG in 1992 and 2002 (both unpublished) were followed by a detailed investigation of the Roman riverfront by ASE in 2008.

Roman Rochester was considered in detail by the VCH in 1932 based on data available then, Harrison and Flight (1968, 75) in the late 1960s and most recently Millett (2007, 166). The town is also well referenced in the Kent HER. It seems likely, based on the excavations of Chaplin (1962, L), that the Roman town of Rochester was located on the site of a pre-existing LIA settlement, though its type has yet to be determined. He reports that during a seven-month winter rescue excavation in 1961 and 1962 on the site of No. 50-54 on the High Street in Rochester (TQ 7427 6863), effectively the centre of the Roman town, what he described as evidence of a Belgic occupation horizon was found below the Roman deposits. The LIA material culture finds included pre-Roman coin moulds, LIA coins, evidence of iron manufacturing, brooches, a gravel trackway and an Iron Age ditch. Chaplin also identified the post-holes of a large wooden structure which he speculated might have been the mint itself given the proximity of the coin moulds.

Some commentators, for example Harrison and Flight (1968, 75), have suggested that this Iron Age site was a fortified settlement (*oppida*). Such an interpretation was based on the later Roman name for the town (see Section 5.1.3) and on Chaplin's finding of the evidence of coin minting which has been interpreted as indicating a settlement of some significance (Chaplin, 1962, L). Everitt (1986, 99) also references Rochester as being an *oppida* prior to the occupation, though clearly these *oppida* interpretations have yet to be definitively proved.

The most visible remnants of Roman Rochester are parts of the stone-built irregular pentangle-shaped city walls dating to around AD 225 which were first outlined by Payne (1895a, 3), parts of which are preserved in the later Medieval walls (at TQ 7431 6849). These are up to 2.2m in thickness with a ragstone and flint rubble core faced with worked ragstone. The stone walls were preceded by a turf wall dating to around AD 175 which featured a large 2.5m deep and 7m wide V-shaped ditch (Harrison and Flight, 1968, 57). The inner face of the later ragstone wall retained part of this earthen bank (covering the stump of the original turf wall) as is evident from the un-weathered nature of surviving wall sections exposed by archaeological excavation.

Rochester is also the likely location of a riverine port given its position as a node for the River Medway and Watling Street, potentially the break-of-bulk point where ragstone was transhipped from *codicaria* to Blackfriars 1-style vessels for onward shipment (see discussion in Section 4.5.3). The ragstone river wall (or potentially an unrecorded bastion) from the occupation period town has certainly been found, by ASE in Horseswash Lane (Jamieson, 2008, 7, at TQ 7424 6884).

The final main feature of Roman Rochester were the roads, with Chaplin's (1962, L) excavations revealing the various phases of development of Watling Street as it travelled through the town and over the bridge. He says:

> *"Conclusive evidence was found of the Roman Watling Street having been constructed as a single trackway with a small side ditch at or about the time of the conquest. This suffered a single remake before being converted into a dual carriageway with a central stone-lined ditch by the addition of a southern carriageway, the new road being at least 6.7m overall. Seven superimposed remakes of the northern carriage-way were noted, the southern one apparently surviving but with two remakes."*

Meanwhile an additional road branched off Watling Street (at TQ 7459 6814), this being the Wealden road which is threaded through much of this book and which passed through modern Maidstone before terminating near the Beauport Park iron working site north of Hastings. Thus, with the Medway crossing and the junction of two major roads, Rochester was a key transport hub during the occupation.

In terms of evidence of the town itself during the occupation, ten definite and two probable stone-built buildings have been found, together with three possible timber buildings (Ward, 2013, 49). One of the stone structures may have been a small temple in a square precinct, located on the high ground now occupied by the Norman castle (Ward, 2013, 49, at TQ 7414 6856). Most recently, a further structure may also have been located beneath the Cathedral during excavations to locate evidence of the pre-Gothic Norman structure (Keevill, 2015, 19). A clay floor surface there has been provisionally identified as Roman, featuring pottery including Samian ware.

Cuxton Possible Villa (at TQ 7098 6648)

Date: 1st through 3rd centuries AD (based on pottery and coin data).

Type: Possible villa settlement.

Site History and Economic Evidence: Cuxton was first investigated as a potential Roman site by Payne and Roach-Smith (Payne, 1902, lxvii) after workmen found 'Roman walls' in the 1890s. Next, in 1950, a Roman cremation burial was found by resident Mr Burrows in Salisbury Row, this being reported in *Archaeologia Cantiana* (Baldwin, 1952, 193). A decade later KAS members P.J. Tester, E.R. Swain and A.C. Harrison carried out the excavation of an occupation-period inhumation burial with the skull placed between the knees (Tester, 1963, 181). Cuxton was considered in detail in the VCH in 1932 based on data available at that time, by Everitt in the 1980s (1986, 100), and is well recorded in the Kent HER. The most significant findings in the area date to the Payne and Roach-Smith investigations, they recording the finding of substantial occupation-period mortared walls in the vicinity of the White Hart Inn (on the modern A228) and the nearby St Michael's Church, together with associated pottery (including some Samian ware), a coin (of Lucius Verus, co-Emperor AD 161 – 169) and Roman roof tiles (Payne, 1902, lxvii, and Taylor, Jessup and Hawkes, 1932, 151). Meanwhile Newman (1969, 253) notes that much re-used Roman material can be found in the structure of the church. This church is of particular interest as, unusually, it is not aligned due east. This misalignment has given rise to a local rhyme: *"If you would see a church miswent, then you must go to Cuxton in Kent."*

The re-used material and unusual alignment have been interpreted as indicating that the original Saxon church was constructed on or near the site of earlier occupation period buildings, perhaps a villa (Taylor, 1932, 111), and Bell (1999, 6) argues that it is this association that has given rise to the name of the village with its *stān* (stone) derivation. He says this is common occurrence in Kent where a Church has an association with a nearby villa site, with other examples including Lullingstone, Stone-by-Faversham and Stone-by-Dartford.

Halling Settlement (at TQ 7045 6319)

Date: Undated.

Type: Possible non-villa settlement/ 'native settlement'.

Economic Evidence: The Roman provenance of Halling was first investigated by Payne (1893, 196) who detailed the finding there of an occupation-period cemetery. He revisited the village in the early 20th century (1902, lxviii) when he reported the finding of a further burial featuring occupation-period burial goods. Most recently ASE conducted a trial excavation in advance of the building of the Halling roundabout on the A228 (Priestley-Bell, 2004, 9, at TQ 7045 6319). Halling was considered in detail in 1932 in the VCH based on data available at that time.

South of Cuxton, Halling is associated in the modern era with chalk quarrying for the now largely defunct Medway Valley cement industry. The most significant finds came to light during Payne's original investigation when he identified a total of ten inhumation burials in the cemetery he located, nine face-down and all with occupation-period grave goods (Payne, 1893, 196). ASE's more recent excavations revealed a potential non-villa settlement with field boundary ditches, post-holes and a further burial (Priestley-Bell, 2004, 10).

Blue Bell Hill Possible Temple (at TQ 7485 6098)

Date: 1st through 4th centuries AD (based on pottery and coin data).

Type: Possible temple.

Site History and Economic Evidence: The remains of a Roman building on Blue Bell Hill were found in 1830 by local antiquarian Thomas Charles who reported the find to Roach-Smith, this being recorded by the latter in the first issue of the RAI's *Archaeological Journal* (1844b, 264). It was later noted by Wright (1854, 177), who also detailed the finding of a hoard of 220 coins. The site was considered in detail by KARU's Kissick (1990, 26), and it features in both the VCH and the Kent HER where it is detailed as a 'temple'.

The location of the building, also marked as a 'temple' on the Ordnance Survey series of maps of the region from 1960 for the next 20 years, is in a small coombe to the east of and above Lower Bluebell Inn on the Rochester - Wealden Roman road (see Figure 21). It was marked as 'Roman remains found AD 1830' on earlier Ordnance Survey maps. Kissick (1990, 26) says that the structure had previously been considered locally to be a watchtower or villa, but by the 1950s a consensus favoured the interpretation as a temple.

Antiquarian finds detailed in the HER and VCH (Taylor, 1932, 110) include Wright's coin hoard (with dates ranging from Claudius to Gratian, Wright, 1854, 177), a brick or tile floor, a nearby 'cemetery', building stone and tile, together with small finds including Samian ware, keys, rings and pins. Kissick (1990, 27) adds that at the time of her research the then owners of the adjacent Blue Bell Hill farm told her that they had found substantial stone foundations beneath their courtyard which predate any known building on the site.

In summary here, there is clearly no definitive identification of the site as a temple, despite it being listed in the HER. Nevertheless, it seems there was a substantial Roman building at this enigmatic site overlooking the vista of the Medway Valley and alongside the Roman road, and in that regard a section of column re-used as garden ornament in Warren Road (at TQ 7490 6170, this being the line of the Roman road) may be evidence of this, although its provenance has yet to be established.

Snodland Villa (at TQ 7075 6202)

Date: 1st through 4th centuries AD (based on pottery and coin data).

Type: Villa settlement.

Site History and Economic Evidence: The first investigation at this site took place in 1844 when Wright (Roach-Smith, 1844c, 164, and Wright, 1854, 186) excavated a Roman barrow burial, in the process exposing a substantial building in Church Field and Stone Grave Field between All Saints Church and the River Medway which he identified as a villa. It was next the subject of observation in 1927 when a new gas works was built in Church Field, with KAS and the Society of Antiquaries funding a trained 'watcher' who recorded a large number of archaeological finds during the construction process (Cook, 1928, 79). Separately, in 1933 a stone sarcophagus was found during the expansion of a local industrial site near to the villa location (at TQ 7075 6200), this containing the skeleton of an adult male (Keith and Ward, 1934, 202). Most recently, in 2009 knowledge of what was by then confirmed as a villa increased considerably when ASE excavated Snodland High Street in advance of re-development of the area (Dawkes, 2009, 1).

The villa site at Snodland is well located, on the banks of the river and between the traditional crossing points of Halling and Aylesford. The initial antiquarian discoveries included substantial walls, tile floor surfaces (some of the latter eroding out of the river bank), tesserae and hypocaust tile (Wright, 1854, 186). One of the buildings was identified at the time as a bath house. Meanwhile the 1927 watching brief revealed a wide variety of material culture finds including a terracotta mask, portrait medallions and coins with a wide date-range from the late 1st to the late 4th centuries AD (Cook, 1928, 80).

The later ASE excavations were on a much larger scale, with the area investigated being located to the immediate west of the known villa site. This work identified that occupation-period activity began in the 1st century AD when a Roman field system (at TQ 7059 6208) was laid out, with a masonry bath house (separate to that detailed above, at TQ7063 6207) constructed between the field system and the villa site. Exposing the southwest corner of the bath house, specific remains included largely robbed-out masonry walls, a portion of hypocaust flue and a contemporary assemblage of ceramic building material. Of the ASE work Dawkes (2009, 41) says:

> *"Whilst rare, bath houses are known from other 1st century civilian sites: a group of palatial courtyard villas on the south coast, including...Angmering in East Sussex and perhaps the best known at Fishbourne in West Sussex. This Flavian bath house is the earliest Roman building to have been identified in Snodland and suggests the contemporary villa belonged to this somewhat select group."*

This bath house survived until the late 3rd century AD when it was demolished and replaced by a larger aisled building which may have incorporated the bath house function. ASE says that two timber buildings to the north and south of the villa were also constructed during this late 3rd century AD rebuilding phase, with the surrounding field system also being reorganised (Dawkes, 2009, 1). The ASE research revealed that by the mid-4th century AD these later masonry and timber buildings had been destroyed by fire. Dawkes (2009, 8) adds that an even later phase of occupation saw the establishment of a small inhumation cemetery and the burying of two coin hoards. In one of the latter some 3,600 coins were found during geotechnical works by the Heritage Conservation Group of Kent County Council (Dawkes, 2009, 1). The site then appears to have then fallen out of use, remaining abandoned until the 11th century AD.

In addition to the coins and ceramic building material, Roman material culture finds from the ASE excavation included pottery, metalwork, domestic items and glass dating from the 1st to the 4th centuries AD (Dawkes, 2009, 42).

Birling Possible Villa (at TQ 6803 6061)

Date: Undated.

Type: Possible villa settlement.

Site History and Economic Evidence: Hasted (1798b, 474) reported that Roman foundations had been found in Oxfield next to the churchyard of All Saints Church in Birling, and since that time building materials of Roman provenance have been regularly found in the plough soil leading to speculation of a villa at this location (Scott, 1993, 102). Tatton-Brown (1996, Birling entry) also reports Roman brick, tile and re-used tufa in the walls of the church. While crop marks of a large structure are evident in aerial photography of the site, no full-scale investigation has been carried out to date.

Wouldham/ Burham Possible Mithraic Temple (at TQ 7138 6250)

Date: 2nd and 3rd century AD (based on pottery and coin data).

Type: Possible temple.

Economic Evidence: The principal occupation-period site here is a possible Mithraeum uncovered in 1893 in a sandbank to the north east of Burham on the bend of the Medway, equidistant between the (then) Wouldham Hall Lime and Cement Works and the West Kent Portland

Cement works during expansion of the latter. It was excavated by Frederick James, then curator of Maidstone Museum, with this investigation being recorded at length in the *Proceedings of the Society of Antiquiries of London* by Payne (1895b, 184) and James (1896, 108), the latter with some fine quality photographs. The structure was considered in detail by Taylor (1932, 110) in the VCH, by Jessup (1956, 168) in *Archaeologia Cantiana* and also features prominently in the Kent HER.

Occupation period activity in the vicinity of Wouldham and Burham should be seen in the context of the significant site at Eccles (below), and also as a potential location for the Claudian river crossing battle of AD 43 (Dio, 1925, 20 and Kaye, 2015b, 239), noting the two modern monuments commemorating this on the eastern bank opposite Snodland church (at TQ 7089 6183).

The possible Mithraeum featured an underground chamber measuring 12m by 6m (it being 4m in height) with three recessed alcoves, a barrel vault roof and an offset zig-zag entrance with high arched entrance (Payne, 1895b, 185). No trace remains of the structure today given the site was ultimately quarried away and subsequently had two lime kilns built on top of it, with the only surviving material being three pieces of chalk wall facing with incised chevrons which reside in Maidstone Museum. At the time of its discovery the site was of considerable local interest, with a major (though ultimately unsuccessful) effort being made to preserve the find and rebuild it elsewhere (Jessup, 1956, 171). The Roman dating is firm given that when the chamber was excavated it was found to contain a Constantinian coin, broken roof tiles, box-flue-tiles and broken pottery of Roman provenance (some identified at the time as being the remains of amphora, Jessup, 1956, 171). Additionally the walls were faced with blocks of local chalk (called 'curly bur') of a type used elsewhere in the Medway valley during the Roman occupation as a building material, they also featuring occupation-period mortar (Taylor, 1932, 108). The antiquarian identification as a Mithraeum was originally based on the alcoves, roof and entrance, but since its discovery there has been much speculation about whether the site was actually an occupation-period cellar rather than a place of worship. In that regard, when examined in detail the chamber was also found to have an adjoining chalk pathway leading to the nearby Medway (where occupation-period wharfing was also found at the time), with a 3m splay identified as a loading ramp. Further, the three niches were then re-identified by Jessup (1956, 170) as being similar to those used for lighting in Gaulish cellars from the Roman period. Based on the balance of evidence Jessup concluded that the chamber was in fact a cellar used to store imported wine and oil, perhaps for the nearby villas at Court Road or Eccles. The case either way for Mithraeum or cellar remains unproven, though Millett (2007, 171) adds a final note of caution when he references the reuse of the cellar at Lullingstone for religious purposes, so perhaps the Burham/ Wouldham cellar may have served a number of functions over the period of its use.

On a final note regarding Burham specifically, St Mary's Church (at TQ 7167 6200, now out of use but maintained by the Church's Conservation Trust) features much re-used Roman building material in its structure, including ashlars of ragstone, chalk and tufa in the external abutments and tile in the nave walls. The likely source is either or both of the Court Road and Eccles villas.

Court Road Villa, Burham (at TQ 7262 6171)

Date: Undated.

Type: Possible villa settlement.

Economic Evidence: Payne (1898, 10) reported the finding of a modest 18m by 10m occupation-period building in 1896 near Burham Court Farm on Court Road which he described as a villa. The excavated site featured box-flue tiles and painted wall plaster, with Payne (1898, 11) identifying a room on the northern side as a bath house.

Eccles Villa (at TQ 7224 6054)

Date: 1st through late 4th centuries AD.

Type: Villa settlement and industrial site (based on pottery, coin and glass data).

Site History and Economic Evidence: The Roman provenance of the villa site at Eccles was first detailed by Poste (1848, 81) who recorded 'sepulchral remains' and coins being found here. It was next revisited by Payne (1898, 12) when the latter considered the Court Road villa site at Burham (detailed above). The Eccles site was then the subject of an aerial photographic survey by LMRAG in 1961 (unrecorded) which revealed the crop marks of a sizeable building, this then leading to test pitting which confirmed a substantial structure at the location. The site was then extensively excavated between 1963 and 1976 by A.P. Detsicas and LMARG, the resulting interim findings being recorded in sequential editions of *Archaeologia Cantiana* from 1964 to 1977 (Volumes 79 to 93, with Shepherd Frere visiting the initial investigations). The Eccles villa site is recorded in detail in the Kent HER.

Detsicas' lengthy investigations revealed the villa here to be one of the largest and finest in Britain (Detsicas, 1967, 170). It shares a number of characteristics with other Medway villas, for example having a very early phase, in the case of this site perhaps as early as AD 55 (Millett, 2007, 152). The principal range of this earliest phase has 12 rooms of which five have high quality tessellated floors, the building also featuring a veranda and a second story (Detsicas, 1967, 173). A large bath house with mosaics, comparable to those at Fishbourne, was built nearby. The quality of this complex has led to speculation that it was owned by a philo-Roman magnate or was the home of a Government official (Millett, 2007, 152), with site finds seeming to indicate a military connection. This initial

Figure 21: Location of the possible Roman Temple on Bluebell Hill, at centre atop the tree line. View from Burham. Simon Elliott.

occupation phase was replaced by a second with marked civilian characteristics, and later by a third and even more grandiose building. As part of this latter range, dating from AD 150 to AD 290, a 60m by 8m outdoor *natatio* swimming pool was added (Millett, 2007, 152) which was longer though narrower than the grand example at the legionary fortress at Caerleon (the latter being 41m by 25m). A final reconstruction after AD 290 turned the villa to face south-west and was of a large courtyard design, this building having at least 37 rooms. The site was still occupied at the end of the 4th century AD, but on a reduced scale. Much resused material from the various phases of the villa estate can be found reused in St Mary's church, Burham (see Figure 22).

Additionally, in 1972 a pit filled with pottery wasters of Roman provenance was found, being interpreted as indicating the presence of a kiln and tilery associated with the villa (Detsicas, 1977, 19, though note Betts also speaks of tile from Eccles being found in London dating between AD 70 and AD 100 (1987, 28).

Allington Possible Villa (at TQ 7510 5782)

Date: Undated.

Type: Possible villa settlement, industrial site.

Site History and Economic Evidence: The Roman provenance of Allington was first noted by Charles (1846, 88) who reported the finding of a Roman building close to Allington Castle. Poste (1849, 65) later reported the

Figure 22: Ragstone, chalk and tufa blocks from the Roman villas at Court Road and Eccles re-used in buttresses of St Mary's Church, Burham. Simon Elliott.

finding in 1847 of a tile-lined occupation-period burial, again near to the castle (at TQ 7499 5772). Further, in 1907 a coin hoard was found in the grounds of the castle (at TQ 7490 5765), unrecorded at the time (Taylor, Jessup and Hawkes, 1932, 144). Most recently a possible kiln has been located by KARU, again near the castle, though this has yet to be recorded (at TQ 7512 5790). Occupation-period Allington was considered in detail in the VCH based on data available at the time and is well recorded in the Kent HER.

Allington is a key location on the Medway as it marks the modern tidal reach of the river and features, moving upstream, the first of the major occupation-period ragstone quarries in the Medway Valley (detailed below in Section 4.3). In terms of the building noted by Charles, the structure found in 1844 comprised foundations and hypocaust pillars. Robertson (1883, 73) later recorded its possible fate, saying that a local had noticed in 1844 masonry from a villa located near to Allington Castle being removed to facilitate ragstone quarrying. Fragments of the hypocaust were saved from the demolition but have since been lost, while Robertson also recorded that tile from the site was also used to mend the approach road to the castle. However, as shown in Figure 23, recent LIDAR imaging seems to indicate that at least some of the foundations of the structure may survive today.

With regard to Poste's tile tomb burial, no further details are available excepting his description of its composition of fire-hardened tiles. Meanwhile, with regard to the hoard, Davies (1982, 137) detailed that it contained 22 coins, one of Claudius Gothicus (Emperor AD 268 – 270), one of Florian (Emperor AD 276), three of the Tetricus II (Gallic Emperor AD 273 – 274) and 17 radiate 'barborous' copies of a variety of Gallic Empire coins (AD 260 – 274). The hoard is generally referred to in the relevant literature (including the Kent HER) as the coin hoard of Tetricus and I stick with that convention here.

Taking all of the above into account, a strong case can be made that the castle (the original structure of which dates back to the 11th century or earlier, though it only became a 'castle' through a licence to crenellate in the 13th century) was built near the site of a Roman villa on this attractive bend on the Medway.

Little Buckland Farm Villa, Maidstone (at TQ 7486 5664)

Date: 1st through 4th centuries AD (based on pottery and coin data).

Type: Villa settlement.

Site History and Economic Evidence: Poste (1858, 156) first recorded the discovery in 1835 near Little Buckland Farm in Maidstone of the foundations of a Roman structure and extensive terrace during the planting of a cherry orchard, the site being interpreted at the time as a villa. Occupation-period pottery and roof tile were later

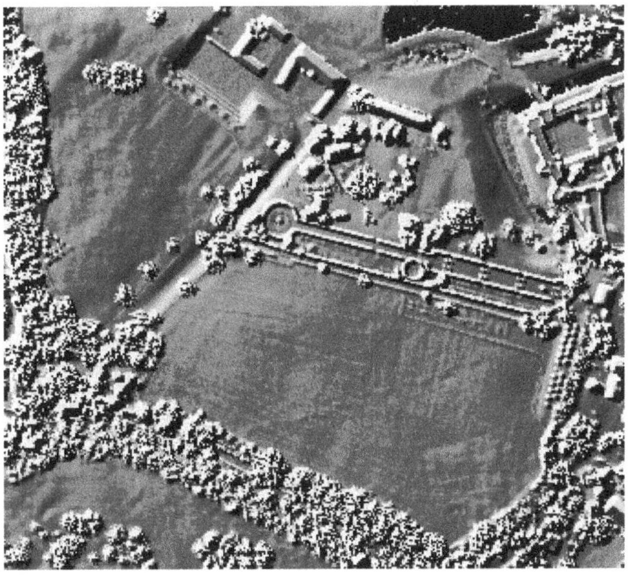

Figure 23: LIDAR image, Allington castle top right, possible villa visible bottom right as a series of rectangular rooms. Supplied by Dr Paul Wilkinson, public domain image, Environment Agency.

found at the site during the building of a sports field in the early 1930s, this being recorded by Wheeler (1932, 99). Most recently a watching brief by CAT in 2003 identified a steep bank close to the site which was interpreted as a small antique ragstone quarry (Goacher, 2012, 12).

Material culture finds from this site include a coin of Constantius II discovered in 1984 (Kelly, 1984, 373) and a bronze figurine of Sylvanus was found nearby in 1820 (Poste, 1858, 168, at TQ 7557 5563).

The Mount Villa, Maidstone (at TQ 7566 5623)

Date: 2nd through 4th centuries AD (based on pottery and coin data).

Type: Villa settlement.

Site History and Economic Evidence: The presence of a Roman villa at this site to the immediate north of modern Maidstone East railway station was revealed in 1843 when a section of the Medway river bank collapsed to reveal Roman foundations and tile. A subsequent excavation by Charles (1847, 86) confirmed the presence of a large structure. It was next the subject of extensive investigation by MAAG through the 1970s (Oldham, 1977, 224), before being definitively excavated by Houliston and CAT in the 1990s (1999, 71, see Figure 24). The Mount villa was considered in detail in the VCH based on data available at that time, and has been the subject of frequent commentary in *Archaeologia Cantiana*. It is also well recorded in the Kent HER.

This villa, sitting proudly on a terrace 8m above the Medway, is one of the best recorded in the region. The first structure on the site was an aisled building dated to the late 2nd century AD which featured a hexagonal water

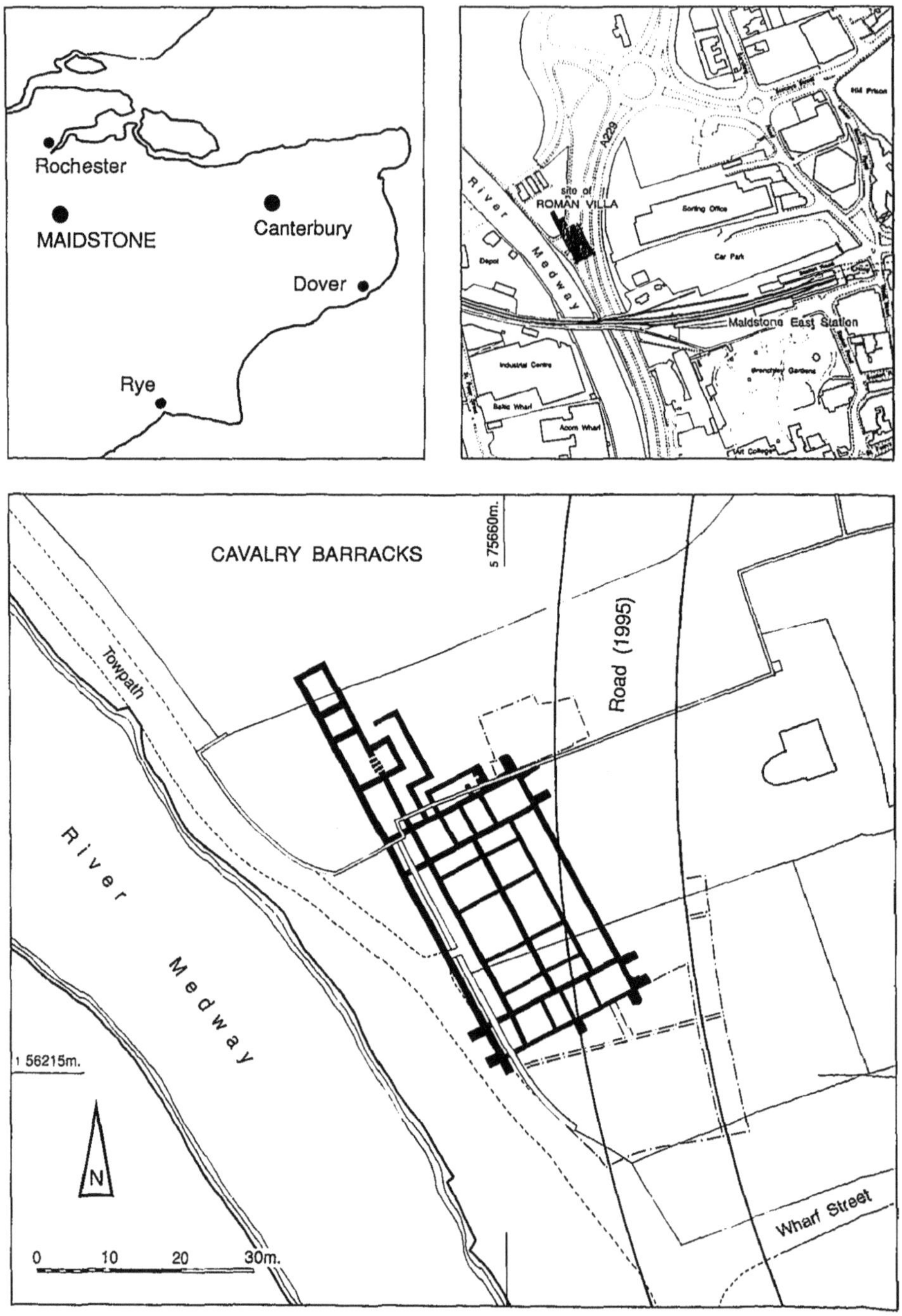

Figure 24: Mark Houliston's (1999, 72) hand drawn plans of the second phase of the Roman villa at The Mount, Maidstone, clearly showing the very close proximity to the River Medway. Archaeologia Cantiana (Kent Archaeological Society).

basin (Blanning, 2014, 181). This was replaced by a large winged-corridor villa featuring a extensive bath house in the early to mid-3rd century AD, this being rebuilt in the early 4th century AD when the site expanded to its largest extent (Houliston, 1999, 73). It then fell out of use in the second quarter of the 4th century AD (Blanning, 2014, 480).

Meanwhile, in the Kent HER and antiquarian record a large number of burials are recorded in the vicinity of this villa, reflecting its location on the Rochester –Wealden road. Further, an occupation-period bronze figure of Mercury was found in 1826 to the immediate east of the villa site (at TQ 7625 565, Poste, 1858, 165).

Finally, on the 1797 Ordnance Survey map of Maidstone a large, elongated island is visible upstream of the Mount site just above the Tonbridge Road bridge. If natural in its origins, the island would certainly have been a feature in the river during the Roman period, but was removed during the early 19th century river improvement programme (it is certainly absent in the Ordnance Survey 1843/44 map). There is a possibility that this island is in fact a man-made feature from the occupation, specifically the result of the digging of a weir-bypass channel of a type described by Ellis Jones' from the Medieval period in the River Severn (2012, 29), these then being known as barge-gutters. Such channels created long, thin islands in the river and, if this interpretation is accurate regarding

Maidstone, would locate an occupation period weir here (see 4.6 below).

Florence Road/ Bower Lane Villa, Maidstone (at TQ 7520 5504)

Date: 2nd through 4th centuries AD (based on pottery and coin data).

Type: Villa settlement.

Site History and Economic Evidence: The third villa site in Maidstone was located in 1893 on the western bank of the Medway at the junction of Florence Road and Bower Lane during drain cutting, though this went unrecorded at the time excepting an unsigned note in the records of Maidstone Museum. It was subsequently the subject of a rescue excavation by CAT in 2004 (Rady and Shand, 2004, 1), with ASE carrying out a further investigation in 2008 (Riccoboni, 2008, 3).

The initial investigation in 1893 revealed the foundations of what was described as a villa, with the later CAT excavations locating in-situ box flue tile and large quantities of both roof and floor tile. The subsequent ASE investigation found a substantial stone floor surface, painted wall plaster, and pottery and coins which date the site to at least the 3rd century AD and most likely earlier (Riccoboni, 2008, 10).

Of particular note, this villa site is immediately above the occupation-period crossing point of the Medway detailed at Tovil below, the latter being linked to Bower Lane to this day. This crossing point, potentially a Roman bridge (again, see below), links the villa location with the Dean Street quarry on the opposite side of the river, suggesting a possible link between elite settlement on the northern bank of the river and industry to the south.

Barton Road Villa, Maidstone (at TQ 7657 5485).

Date: Undated.

Type: Villa settlement.

Economic Evidence: This site was first excavated in 1870 by Hubert Bensted on behalf of KAS, this investigation recorded by Roach-Smith (1876, 163) who included a detailed floor plan. It was further investigated in 1929 during the building of Maidstone Boys Grammar School though this work went unrecorded. The site is considered in detail in the VCH.

This villa site sits above Maidstone on the line of the Rochester - Wealden Roman road as it departs modern Maidstone and heads south. The building, of a large courtyard design constructed largely of ragstone (with tufa also being used), featured hypocaust tile pillars, a tessellated pavement of geometric design and an octagonal room thought to be part of a bath house (Roach-Smith, 1876, 163). Given its positioning alongside the Wealden road, a reasonable case can be made that this, or another nearby site yet to be located, was a *mansio*. However, when discussing the subject Ellis Jones (2012, 29) points out that such way stations were usually between 40km and 56km apart. In this context, and taking into account Rochester's positioning on principal Kentish thoroughfare Watling Street (and indeed the Medway), the latter would perhaps have been a more likely location for such a facility.

Boughton Monchelsea 'Remote Bath House' (at TQ 7792 5153)

Date: 1st through 4th centuries AD (based on pottery and coin data).

Type: Bath house and industrial site.

Site History and Economic Evidence: This enigmatic structure was first discovered in 1841 in a field called The Slade to the immediate east of the Roman quarry here (see 4.3) by local antiquarian Clement Taylor Smythe, he reporting the find to Roach-Smith who then recorded it in *Archaeologia* (1842, 414, this including a detailed floor plan). The site was considered in detail in the VCH, and again by Blanning (2008, 7).

Initial settlement in the Boughton Monchelsea area is associated with an extensive LIA earthwork which Howell (2014, 38) argues was an *oppidum* based on its size, similarity to already identified *oppida* in northern Gaul and its setting on a slope as opposed to a hilltop (which he says would indicate a hillfort). The earthen banks at the site are still visible and accessible (Everitt, 1986, 100, at TQ 7656 5158).

However the remote bath house in The Slade is the best known site here. As discovered in 1841, the 18m long structure comprised five rooms (which included three apses) featuring hypocaust tile, window glass, painted wall plaster and cemented basins. The earliest coins found in a primary context date to the 1st century AD, with the latest from the reign of Valens (Emperor AD 364 – 378). Blanning (2008, 7) explains that the bath house is one of four in Kent which show no signs of any accompanying structure (the others being at Baston Manor, Kemsing and Little Chart), she reasonably arguing that it was a meeting place for local businessmen (*negiatores*), possibly associated with the quarrying industry, lying as it does along the same Roman road as the Lockham Wood walled cemetery detailed below (Howell, 2014, 58). Hastings (2000, 10) also reflects on its remoteness, though he arguing that it was the Roman equivalent of a 'pit head' bath for the skilled quarry workers. Either way, the link with ragstone quarrying seems logical given that Boughton Monchelsea is the site of the first of the four major occupation-period quarries upriver of Allington on the Medway and its tributaries (see Section 4.3).

As with the other occupation-period quarry sites, Boughton Monchelsea and its environs also feature a variety of Roman settlements and associated infrastructure in addition to the bath house. Scott (1993, 103) details a potential villa at TQ 7835 5166 with finds including tile, brick and pottery, though specific data to warrant a separate entry is lacking. Additionally, the foundations of two 2nd century AD buildings have also been found within the boundaries of the *oppidum*, one utilising ragstone and the other flint and associated with the nearby quarrying activities (Howell, 2014, 61, at TQ 7832 5168). Further, to the immediate west at Pimpe's Court, the foundations of two other Roman buildings are detailed in the HER together with associated material culture including coins and Samian ware (at TQ 7489 5255). Finally, at the point where the original route of the Roman - Wealden road crosses the Loose Stream, there is a stone marker which is likely of Roman provenance given the use of *tegulae* in its construction. Further investigation of this latter is planned.

Boughton Monchelsea 'Lockham Wood' Walled Cemetery (at TQ 7765 5220)

Date: 1st to 3rd centuries AD (based on pottery and coin data).

Type: Cemetery.

Site History and Economic Evidence: The occupation-period walled cemetery at Lockham Wood, close to the remote bath house site, the nearby Roman quarry at Boughton Monchelsea and 10m from the Rochester – Wealden Roman road, was first excavated in 1842 by Smythe who detailed substantial funerary activity in a 26m by 24m ragstone-walled enclosure. His original manuscript recording his extensive investigation, prepared for the Society of Antiquaries, went unpublished during his lifetime and was finally published posthumously in *Archaeologia Cantiana* (1883, 81). An attempt to re-find the site in 1996 by SEAS, based on its exact location as listed in the Kent HER, failed although evidence of Roman activity was noted (Gardiner, 1996, 5). A similar attempt by the Museum of London Archaeology Service (MOLAS) in 1998 also failed, though this investigation did find evidence of occupation-period ditches associated with Roman field systems alongside the road (Mason, 1998, 15). Finally, in 2000, MOLAS relocated the walled cemetery together with four additional cremation burials (some with grave goods) outside the walled enclosure (Neilson, 2000, 33).

As found by Smythe (1883, 83) the walled cemetery included two 'tower' mausolea, one rectangular (measuring 4.3m by 3.6m) and one circular (with a diameter of 3.5m), together with seven cremation burials. Two of the latter were in cists, with some re-using amphora. Substantial material culture artefacts associated with the burials were also found across the site including pottery, glass, coins, bronze vessels and iron lamps, suggesting that it had remained in use from the 1st to the 3rd centuries AD.

The later MOLAS investigations also found further field system ditches, providing additional insight into how the walled cemetery fitted into the wider pattern of local land use at the time (Neilson, 2000, 33).

Reflecting on the cemetery using the originally available data, Philp (1968, 6) believed it had very close parallels with that at Keston, including the association of a circular and a square tomb with other cremation burials.

Tovil

Date: Undated.

Type: Possible river crossing, weir and wharf.

Site History and Economic Evidence: Tovil, the first village on the River Medway upstream of Maidstone, has no published Roman provenance outside of the mention in the Kent VCH of burials associated with Roman pottery being found (Wheeler, 1932, 101). However it has become the focus of significant research by the author given that it marks the point where the huge Dean Street Roman quarry (see Section 4.3) meets the river. Specifically, below I detail three investigations, firstly a possible Roman river crossing, next an antique weir and finally the 'Medway Stones'.

Regarding the river crossing, this site is now the location of a modern footbridge across the Medway at TQ 7520 5486, with Smith (1839, 56) having detailed the same spot as an ancient crossing point. Most significantly however, examination of the 1797 OS map of Maidstone shows two very similar oval islands sitting parallel in the river at this crossing point. The most northerly was removed by the Medway Navigation Company at some stage before the publication of the 1st edition 25" (1862-75) OS map of the area, while the southern island was incorporated into the southern bank by the time of the 2nd edition (1897-1900). The latter now marks the southern footing point of the modern footbridge. These islands, as recorded intact on the 1797 OS map, have a very interesting configuration given that they are aligned with the 'ancient crossing point', and significantly align to the north along modern Bower Lane to the site of the Florence Road/ Bower Lane villa site as it sits above the river. Meanwhile, to the south the islands also align with the exact point where the Dean Street quarry meets the river. A hypothesis can therefore be advanced that the islands are actually the remains of bridge piers for an occupation-period crossing linking elite settlement on the northern bank with industrial activity on the southern bank. This correlation of the villa on one bank and the quarry on the other is strong evidence that this location did indeed feature a Roman bridge. A clear analogy can be seen in Rome where bridges across the Tiber were specifically built to link elite settlement on one bank with industrial activity on the other, for example the *Pons Aelivs* bridge linking an elite site on one bank with the Marmorata marble working yard on the other, and similarly the *Pons Agrippae.* Intriguingly both of these bridges feature two bridge piers. In the case of a Roman bridge spanning the Medway at Tovil, this would of course have required a drawbridge to facilitate riverine navigation, as with the Roman bridge at Rochester (two in the case of the latter).

Meanwhile, on the same 1797 OS map a weir is also visible some 20m upriver of the Tovil crossing point, this also now removed, though an associated section of ragstone river wall is still visible when the river levels are very low. It is unclear if this marks the site of Roman river infrastructure (interestingly it does appear to have a tile-bonding layer), and again further examination is planned.

Finally there are the 'Medway Stones', found in the river between Tovil and East Farleigh at grid reference TQ74099 53918. This story begins with the EA alerting the author in May 2014 that four large 1.5m diameter circular stones had been dredged out of the river mid-channel during annual river maintenance (see Figure 25). On close examination it was immediately apparent that they had been worked, with the dredging team explaining that they believed there was a spread of many more over a range of 30m around the find spot on the riverbed. The author recorded the four retrieved stones and then engaged the academic archaeological community to gather insight into what the stones might represent. The result was a wide range of views, with the most common interpretation being that they are blanks for millstones (not querns given their size using Shaffrey's definition, 2015, 78), or part-finished Roman columnal bases.

As detailed above in Section 2.5, one of the stones was petrologically examined shortly after they were found and identified as a type of Greensand, they appearing to have actually originated in the Folkestone region and been part of a cargo on its way either to the Medway Valley or elsewhere but with the carrying vessels stopping off at this location (presumably to pick up a load of ragstone).

Later in 2014 the author invited the Royal Engineers based at Invicta Barracks in Maidstone to carry out an exploratory

Figure 25: Initial 'Medway Stones' as recovered by the Environment Agency before recording. Simon Elliott.

dive on the site, this being undertaken in October that year (see Figure 27). The dive confirmed the wide spread of material on the river bed, including more of the circular stones (one piece of potential millstone is visible from the bank at low tide), possible ships timber and most excitingly a row of piles driven into the riverbed some 3m out into the river and running over 20m downriver (see Figure 26). This reflects the similar experience at the

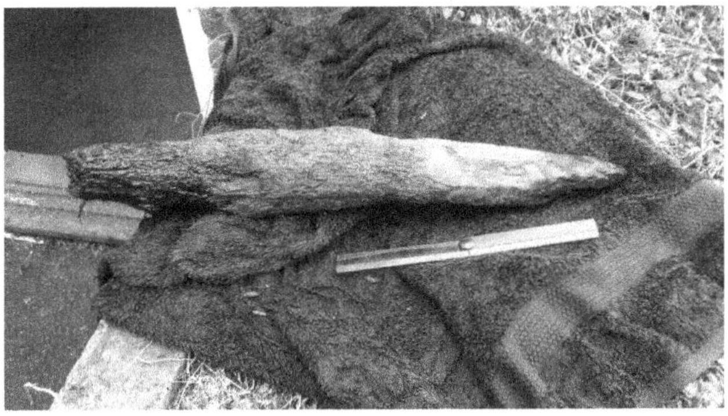

Figure 26: Wooden pile stump recovered from the River Medway at the point where the 'Medway Stones' were found. One of a row running for 20m, set 3m out into the river. Simon Elliott.

Figure 27: Royal Engineers from Invicta Barracks in 2015 conducting an investigative dive on the 'Medway Stones' wreck site at the request of the author, under the supervision of the author and KAFS. Simon Elliott.

recent examination of riverine infrastructure in the River Itchen in Southampton where Russel (2013, 7) describes the identification of 179 piles in two piers and a revetment dating to the early 3rd century AD which, significantly, are also set out into the river. Providing possible insight into the positioning of the Medway piles, especially given that the Itchen piles are located close to sites associated with the *Classis Britannica* (see discussion in Chapter 6), he says (2013, 7): *"...it is possible that the piers were always well out from the west bank to access the main channel that passes much closer to the east bank."*

The implications of these finds, if the provenance can be established, are profound. This is both in terms of pure Roman archaeology, but also more broadly with regard to the experiences of those working in and facilitating the extractive industries during the Roman occupation. Firstly, it would indicate the location of a wreck in the river of a Roman vessel similar to the Blackfriars 1 ship, set against a wharf at a point close to where the Dean Street quarry meets the Medway. To get an idea of the size of industrial river structures of this type, the 3rd century AD quay at the legionary fortress of Caerleon was built to an impressive height of 6.65m above sea level (Boon, 1978, 1) while the waterfront buildings associated with the wharfing at Lincoln on the River Witham extended for a kilometre (Ellis Jones, 2012, 78). Secondly, confirmation of a Roman wreck would provide insight into the industrial processes being applied in occupation-period quarrying, for example confirming that blanks were being prepared on site in their quarry of origin to order, thus ensuring that best use was made of each boat journey with no cargo space being wasted.

Gallants Lane Iron Working Site (at TQ 7270 5360)

Date: Undated.

Type: Possible industrial site.

Site History and Economic Evidence: This site was first located in late 2015 by the author. During regular walkover surveying of the area it became clear that there was a dense concentration of iron working detritus including tap slag (see Figure 28), cinder and charcoal on the southern slope of the Medway Valley between the Dean Street and Quarry Wood Roman quarries (see Section 4.2). Within this concentration was also found broken up Roman roof tile and occupation-period pottery. Detailed examination of the area on google earth and aerial photographs has revealed a dark crop mark at this specific location, together with many others up the slope towards the high ground of Coxheath above the river valley. Walkover surveying of two of these additional sites has revealed more iron working detritus, while in an adjacent antique tree line untouched cinder heaps have been found in association with the ragstone foundations of a possible road surface. The location of all of these sites may therefore indicate a chronological sequence of occupation-period iron bloomeries associated with the nearby quarries, manufacturing and repairing tools for this industry. The positioning of such metal working sites to support nearby industrial activity are a common feature of the Roman industrial landscape, with for example Carroll (2016, 31) highlighting the presence of such an operation next to the extensive Vagnari wine-manufacturing Imperial Estate in Italy.

Figure 28: Tap slag found by the author walkover surveying the Gallant's Lane possible iron working site, the first such if proved of Roman provenance outside of the Weald in Kent. Simon Elliott.

Siderite to facilitate the iron manufacturing at the Gallants Lane site, if proved, could either have been transported in by road from the nearby Weald, or quarried locally from the exposed seam of the Wealden Clay which sits beneath the Hythe Beds (the source of the local ragstone). With regard to the latter option, a small extinct quarry does sit at the foot of the slope where the iron working sites have been located, clearly being extinct on the 1797 OS map of Addington.

Most recently the wider metal working provenance of this site has been further bolstered by the finding of a 15cm elliptical, bowl-shaped lead alloy ingot (120cm in diameter and weighing 1.71kg) in a neighbour's garden at TQ 73374 52672, its occupation-period origins now being the subject of metallurgical analysis.

East and West Farleigh Villa, Temple (at TQ 7270 5360)

Date: 1st to early 5th centuries AD (based on pottery and coin data).

Type: Villa settlement with temple.

Site History and Economic Evidence: The foundations of a Roman structure were originally discovered here in 1838, this being reported by Smith who also provided a plan (1839, 57, though with no references giving any context). He also detailed the finding here of a coin of Florianus (Emperor in AD 276). Additional foundations were found in 1938 during the cutting of a bank across the site, this being reported in *Archaeologia Cantiana* (Fisher, 1939, 204) though no further investigation was possible at that time. MAAG then returned to the site in 2005 and have conducted annual investigations every year since

Figure 29: Site of the villa at East Farleigh on the south bank of the River Medway, in the green field behind the brick building at centre. Simon Elliott.

(see Figure 29 for location). Site Director Albert Daniels has periodically reported on progress in the CKA's *Kent Archaeological Review* and the *KAS Newsletter*.

The earliest evidence of human activity in the Farleighs is a substantial LIA ditch at the villa site detailed above, it being interpreted as defensive in nature due to its size and very steep sides (Daniels, 212, 179). The LIA provenance of this initial occupation has recently been further reinforced by the finding of a silver LIA *minim* attributed to Tasciovanus, a north Thames chieftain.

The next phase of activity is evidenced by an early occupation period ditch of even greater proportions, this being 2m in depth and 3m wide, with the backfill used as a bank to the south to form an obstacle 4m in width (Daniels, 2015, 6). This latter ditch then runs beneath at least three phases of settlement activity lasting for the entirety of the occupation, before the whole area was abandoned until the advent of the Anglo-Saxon churches at East and West Farleigh.

The earliest phase of Roman structural occupation is one of four buildings found in the 2007 season by MAAG which dates at the latest to the early 2nd century. Dubbed Building 2 based on the chronology of its finding, it featured a grey clay floor surface into which stones had been pushed to form a metalled surface. The building also had associated ditches, it being subsequently demolished with the next range of buildings constructed on a slightly different alignment (Daniels, 2015, 6).

This next phase of activity was identified in the initial 2005 season when a building of 29m by 15m with three rooms and featuring on three sides a 4m-wide corridor was uncovered, at right angles to the later buildings. Dated to the 3rd century AD and dubbed Building 1, this is thought to be that recorded by Smith in 1839. The three other structures found in 2007 also date to this period of occupation, including a Romano-Celtic temple dubbed Building 5 (see Figure 30). The latter building has been 0fully excavated and measures 13m by 11.5m, consisting of two rooms on three sides of which is a 2.5m-wide corridor. The walls are mortared ragstone which in places remain over a metre high and feature fragments of painted (to represent marble) and plain wall plaster on the exterior face, while some of the floors are finely mortared with

Figure 30: Building 5 at the East Farleigh villa site under excavation. Columnal bases for monumental entrance in foreground, facing towards the River Medway. Likely place of origin of monumental arch at All Saint's, West Farleigh. Simon Elliott.

rendering around the base of the walls. The walls of the two inner rooms are substantial and appear to be the base of a two-storey central structure, with the surrounding corridor being single storey in design (and in a north western European context being enclosed rather than colonnaded, hence the external painted wall plaster). The temple also features a grand engaged column doorway facing the river, though the building itself is orientated east-west to face the rising sun. In this regard it is both a typical 'hello-goodbye' temple for offerings to be made before and after a river journey, hence the main entrance facing the river, but also a place of worship for more traditional classical deities.

By far the most high profile development regarding this temple was the finding of a 6cm long lead curse scroll against the north west corner in a demolition layer in 2009 (English, 2012, 3). Dated to the 3rd or 4th century AD, this was carefully unrolled and studied under a scanning electron microscope, with Dr Roger Tomlin at Wolfson College Oxford revealing 14 personal names written in capitals in two columns. Some of these were backwards or upside down, possibly to emphasise the effect of the 'magic' of the scroll on these individuals. Specific Latin names revealed include *Constitutus, Sacratus, Constans* and *Memorianus*, while intriguingly two of the names are indigenously British or Gallic (*Atrectus* and *Atidenus*, English, 2012, 4). Meanwhile, pottery featuring graffiti has also been found around the site and this is currently being deciphered to see if it is also of a religious nature.

Based on the dating of the scroll the temple seems to have been in use for religious purposes late into the occupation, though the building was definitely being differentially re-used by the end of the 4th century AD given two circular ovens and a pair of quern stones were installed in the two central rooms, indicating the structure's use as a bakery at this late date. The querns are of particular interest as the upper stone is made of Greensand sourced from the Folkestone area while the lower is of millstone grit sourced from Derbyshire or Yorkshire, indicating that long-range trade was still in operation in this area at this late date.

The evident reuse of Building 5 as a bakery represents the third and final stage of Roman activity on the wider site. Similar late reuse also appears in Building 3 (thought to be part of a barn complex), which seems to have had an entrance with a layer of broken roof tile capped with clay inserted as a cart access. Daniels (2015, 6) explains that additionally, in the western part of this building, a circular structure 1.2m in diameter (featuring a 200kg piece of re-used ragstone) with a 200mm wide flue to the north has been found which he argues was a corn-drying oven associated with the Building 5 bakery.

In terms of dating the end of occupation at this site, the latest coin finds are one of Honorius (a pre-AD 402 issue) and four of Arcadius, indicating that the site fell out of use by the late 4th or early 5th century AD. This is backed up by a recent coin hoard found comprising fifty coins dating to between AD 350 and AD 365 (Daniels (2015, 7). The whole site then appears to have been totally abandoned for centuries after the occupation, with today's visible remains being covered over by hill wash, hence their excellent preservation.

Some of the building material from this villa appears to have been re-used locally following the occupation. Both St Mary's church in East Farleigh (at TQ 7342 5331) and All Saints church in West Farleigh (at TQ 7157 5350) feature re-used Roman tufa in their original Anglo-Saxon structures. In the case of the former this is in the form of ashlar blocks built into the first Church tower on the site, now embedded (though still visible) in the later ragstone Norman tower (see Figure 31). The tufa stonework, classically Anglo-Saxon long and short work (Godfrey, 1962, 365), shows evident saw marks from the original blocks being cut to fit the later requirement, with closer examination revealing Lewis holes from the original Roman construction process, beam slots to support a

Figure 31: Re-used tufa from the East Farleigh villa site in St Mary's Church, East Farleigh, from the tower of the original Saxon church. Simon Elliott.

roof structure and *spolia* (in this case re-used engaged columns).

Meanwhile, All Saints church exhibits even finer re-used Roman tufa in the structure of the building's original Anglo-Saxon nave, specifically in the form of a complete grand arch, a second doorway and in window frames (Elliott, 2014a, 253, see Figure 32). The large east-facing arch, the original main entrance to the Anglo-Saxon church, is 275cm across and appears to be the re-used main entrance to the Building 5 temple at the East Farleigh site which is the same width and features tufa capital columnal bases with the arch removed. Likewise the second tufa doorway in the church (at the opposite end of the nave to the arch) also seems to have originated from the temple site and indeed is of a similar size at 127cm to the Roman structure's side-entrance. Other locally re-used material from the villa includes ragstone blocks found in the garden wall of the Old Vicarage on Lower Road, close to the villa site, which feature chevron patterns exactly like those found on the chalk blocks from the Wouldham/Burham 'mithraeum'.

Figure 32: Monumental tufa arch showing heavy weathering on the original external and internal faces, All Saints Church, West Farleigh. Likely re-used from the East Farleigh villa site. Simon Elliott.

A second East Farleigh villa is also recorded in the antiquarian record and on the 25" OS map from the 2nd Edition onwards (at TQ 7284 5388), it being detailed by Payne (1880, 168) in *Archaeologia Cantiana* (see Figure 33 for likely location today). He says that in 1879, during market gardening on the northern bank of the Medway (immediately opposite the villa above), a tile-paved cistern (with regular tiles measuring 40cm by 30cm) was found made of mortared ragstone and tufa with a layer of bonding tile. It was full of broken roof tile and building detritus and was dubbed at the time a water tank. Later excavations revealed two substantial walls of ragstone and tufa and another tiled floor surface, the whole site being covered in broken roof tile indicating that the roof had collapsed after the building was abandoned. Payne also said that this new site had its own private burial ground, though he provides no details other than to say that it was known locally as the 'lesser cemetery', presumably in contrast to the walled cemetery at Barming detailed below.

The only dating evidence available from the 19th century excavations was a coin of Titus (Emperor AD 79 to 81) found at the site. MAAG attempted in the 1960s and 1970s to relocate this villa but was unsuccessful, except for finding neatly stacked Roman roof tiles in nearby hedgerows, which dated from the original 19th century excavations. Given that the original finds were made at a depth of over 2m after double-depth digging by market gardeners, it is perhaps not surprising that this relocation attempt was unsuccessful. However, most recently, unpublished research by members of the Farleighs History Society has found a copy of the 1st Edition OS map in Maidstone Museum which has in pencil an arrow and

Figure 33: Site of the second at East Farleigh on the north bank of the River Medway. Simon Elliott.

notes illustrating the actual find site. While the provenance of this is still being ascertained, the location does seem a good candidate as on the 1st Edition the specific area is shown as wooded, while in the 2nd it is a fruit orchard. This may indicate the market gardening process between the two OS editions which facilitated the site being found. Once again, further investigation is planned.

We can next consider the river crossing itself. As fully detailed in 4.5.2 below, Coles (1630, 134, see 5.5.3 below) notes the finding in the Medway in the 17th century of: *"A (stone) shelf at the head of an island, many trees on the island, to the north, in default of Augustine Skinner, and Robert Hodges."*

The only islands in the geographical reach of the river covered by this document and known at the time would have either been the two at Tovil or one at East Farleigh, the latter visible on the 1797 ordnance survey map to the immediate left of the bridge. Given that Augustine Skinner lived in West Farleigh (being the local Member of Parliament, elected in the Long Parliament in 1642), a good case can be made that the island described by Coles is the nearer one at East Farleigh. The removal of this 'stone shelf' would have produced a large quantity of locally usable good quality building material and indeed slabs of heavily marine weathered stone can be found around East Farleigh, for example as capping stones on St Mary's East Farleigh churchyard wall.

Finally for East Farleigh we can consider a number of sepulchral burials from the antiquarian record. One group was found in 1845 adjacent to Gallants Lane (at TQ 7284 5238) along the line of the Roman road running from the Dean Street quarry (or indeed as a spur running from the Rochester-Wealden road further east) to the Barming ford and East Farleigh villa. This consisted of cremation urns in a stone cist, being dated to the later 2nd century AD through coin evidence. Saynor (1974, 2) detailed that the cist featured at least four urns containing cremation burials (one in a Patch Grove storage jar, another in a jar thought to be of similar provenance and two in re-used and unidentified flagons), together with a small ornamented bronzed castor vessel and an empty black globular vessel thought to be made of glass. All the urns had had their rims removed, either deliberately or by ploughing. More recently, and intriguingly, the author has found 30m along the same Roman road the rim of a very large Gauloise 4 amphora (at TQ 7238 5273) dating the late 1st century AD. This could be another as yet unrecorded cremation burial, re-using the amphora. This is unusual for two reasons, firstly because there are only two other cremation burials housed within amphora known in western Kent (and both of these at the extremities of the county, Blanning, 2014, 35), and secondly because of the type of amphora given that most amphora cremation burials in Kent use Dressel 20 types. A useful regional analogy would be the unusual Colchester St John's Street cremation burial (Symonds and Wade, 1989, 85) excavated in 1986 in which a Gauloise 4 was found intentionally split in two pieces halfway down its length to enable a large jar (containing the human remains) to be inserted into its neck. Even more intriguingly at this site, the rim was found while the author was investigating what was originally thought to be a Roman milestone on this road at the same location, but which may now be considered as a burial marker.

Meanwhile, an additional burial was found in 1843 in the Old Vicarage on Lower Road (at TQ 7308 5330), while in 1797 the Kent HER details seven burial urns found to the immediate north of the 'water tank' villa on the northern bank of the Medway (exact site unrecorded).

Barming Villa (at TQ 7205 5412)

Date: Undated.

Type: Villa settlement.

Site History and Economic Evidence: This villa site was first recorded in 1797 by the then rector of St Margaret's Church, antiquarian and biographer Rev. Mark Noble (as detailed above in 4.1.1, unpublished at the time and first recorded by Roach-Smith, 1848, 183). Walls of the villa were still visible into the 19th century, with Smith (1839, 59) confirming that: *"Till within a few years ago the remains of a Roman Villa and bath might be seen in the meadow between (Barming) church and the river."*

Subsequent attempts to locate the site were unsuccessful and it appears that all evidence of its existence was cleared away by local quarrying activity before 1848. Attempts in the 1970s by MAAG to locate the villa only found occupation-period building debris in the plough soil of a field to the immediate east of the likely villa location. This appears to have been the dump for the early 19th century quarrymen when they hacked through the villa site to access the ragstone beneath. Most recently a rescue excavation in 2007 by MAAG in St Margaret's Church car park, in which the author participated, only found a few fragments of Roman roof tile (Daniels, 2008, 1).

The main range of this villa as located in 1797 was to the immediate south west of St Margaret's churchyard, it featuring a number of small rooms and hypocaust box flue tile. In association with the villa site, Noble also found a walled Roman cemetery the same year, immediately to the north of Church Lane which links South Street to the church. It is still visible as a crop mark on the December 1990 Google Earth image of the location (at TQ 7227 5425). Such a cemetery would have been located on the approach road linking the villa to any nearby roadways. The cemetery is famous in the antiquarian record for a Roman sarcophagus lid, broken into three pieces, which was found just outside the north wall. Each of the pieces was decorated with palm-tree patterns, interpreted at the time of their finding in 1848 as representing a Christian burial (Taylor, Jessup and Hawkes, 1932, 145). The lid pieces, once stacked in the Churchyard, have since been lost. The site as a whole is very reminiscent of the walled cemetery detailed above at Lockham Wood. Most recently, finely worked building ragstone has been found on the northern bank of the River Medway by the author, immediately to the south of the villa site.

Barming is also one of the historic crossing points on the upper Medway (and the point where the Roman road from the Dean Street Quarry meets the river, at TQ 7240 5391), today featuring a footbridge which locals believe is built on the cobbles of a Roman ford visible when the river level is very low. It was while carrying out desk research in pursuit of this crossing in the historical record that I came across the Coles references (see above regarding East Farleigh and Section 4.3.3) referring to stone riverine infrastructure being removed from the River Medway in the 17th century. In addition to referencing East Farleigh, Coles also notes (1630, 134) the removal of: *"...a foundation of a weir and many stones and Hafsocks and a (stone) shelf at St Ellen's...."*

St Helen's is an historic reference to Barming, the name of the local manor and the name of the street from Lower Road in East Farleigh to the current Barming footbridge. Given that this is even further upriver than East Farleigh, and far removed from the navigable Medway until the modern era, and finally that there is also no nearby recorded Medieval religious institution, there appears little explanation for a weir foundation here unless it dates back to the occupation. Significantly, Hasted (1797, 2) reports that the original bridge over the Medway at Barming (destroyed in 1795) was built by the same Commissioners of Sewers whom Coles reports as removing the weir above.

Teston Villa (at TQ 6993 5314)

Date: 1st to early 5th centuries AD (based on pottery and coin data).

Type: Villa settlement.

Site History and Economic Evidence: This grandiose villa in the upper Medway Valley was first located in 1872 by hop-field owner Arthur Fremling who excavated the bath house. The investigation generated much regional interest, with the Rev. J.M. Mayhew presenting the findings locally and the site then being recorded by Grover (1873, 45) who also provided a detailed plan of the findings. The site was from that time detailed on the OS map series, but the location was misrecorded and subsequent attempts by MAAG to find it in the late 1970s failed. It was eventually relocated by CAT in 1991 when carrying out a trial excavation on behalf of Southern Water whose waste-water treatment works runs along the western boundary of the property on which the villa is located. CAT uncovered four walls, three of which had been robbed out and one of which featured a thin coat of painted wall plaster on the remaining ragstone courses (Rady, 1992, 7). The site was next the subject of investigation after the current site owners attended a talk on the Roman Medway Valley by the author in early 2013, they presenting a roof tile featuring a hand print (Elliott, 2013, 40). A subsequent walkover survey revealed the Roman structure to be eroding out of the slope above the Medway, this being confirmed by a field walking exercise by KAFS who then began an ongoing annual excavation of the site. The Teston site has also been the subject of geophysical surveying by Staveley (2012). The villa at Teston was considered in detail by the VCH based on the data available at that time, and most recently by Elliott (2013, 40). It should be noted that although this site has been the subject of the numerous investigations detailed above, it has yet to be officially documented and so is considered holistically here for the first time. An Interim Report on the current investigations, together with floor plan, is scheduled for publication in early 2018.

The site at Teston features the furthest upriver substantial evidence of occupation-period activity on the River Medway (though see 3.1.4 above regarding Great Cansiron in the Weald, and the discussion regarding East Peckham below). The principal structure is a substantial villa on the north bank of the Medway, with at least two phases of occupation. The main range of the first, a large though simple aisled hall design (Elliott, 2013, 44), was comprehensively demolished in the 2nd century AD (based on pottery and coin data). Surviving wall plaster, box flue tile and a substantial doorway indicate a grand affair for the early occupation period. Its demise may be linked to the geology of the site, which while luxurious in terms of fine living on a south-facing slope above the Medway in verdant countryside, is prone to the emergence of springs when the local water aquifers are full after extreme weather events.

Of the second phase the remains are much more substantial, the winged-corridor design being located further up the valley slope than the first main range and away from the proliferation of springs evident today. It was the bath house of this building that Fremling excavated in 1872, it featuring an extensive hypocaust system with substantial quantities of box flue tiles together with in-situ hypocaust tile pillars. An attached apsidal building with built in drainage was also found at the time, this being interpreted as a plunge pool.

The most recent excavations by KAFS have uncovered most of this second phase of occupation, re-finding the bath house (see Figure 38 for associated hypocaust channel) and then revealing the largely intact foundations of the main range of the building (up to 64m in length) together with one of the wings which had foundations substantial enough to support three stories (Elliott, 2013, 43). The 2015 excavations additionally revealed a substantial multi-room range of buildings immediately adjacent to the main range which appears to have been a kitchen range given the presence of burnt areas interpreted as ovens. The earlier geophysical survey (Staveley, 2012) also revealed the location of the unexcavated wing (this also featuring an apsidal, indicating either a fine dining room overlooking the River Medway or an additional bath house). The survey also revealed landscaping down the

Figure 34: River-front along the Medway immediately below the Teston villa site. The width of the river today for navigation (with the aid of locks and weirs) is very evident. Roman brick and tile has been found in the northern bank by the author at right. Simon Elliott.

slope of the valley towards the river, this being confirmed by test pitting which exposed substantial retaining walls.

Material culture finds at the site (Elliott, 2013, 44) include a tin alloy disc brooch (with two circular registers of enamel, the outer with 19 small panels of fine millefiori glass with space for five more, see Figure 35), copious amounts of window and decorative glass, pottery from across the occupation period (including Samian ware in the first phase of occupation), broken amphora, spindle whorls, a bone pin and a wide range of coins (found in both primary and secondary contexts). The earliest coin from the first phase of occupation is a *denarius* of Nerva (AD 96-98), while in the later excavated building plentiful bronzes of Constantine I (AD 330 - 337), Constantine II (AD 337 - 340), Vetranio (AD 350) and a single coin of Honorius (AD 393-423) have been found. Over the two phases of occupation this is a very wide range and parallels the experience at nearby East Farleigh. Of particular interest, in the very latest occupation layer of the second building phase, recycling hearths for glass and lead have been found dating to the late 4th/ early 5th century AD, together with neatly stacked roof tile and pottery of early Germanic Saxon style. The former two indicate that an attempt was made to recycle the building materials from the villa when it fell out of use in the late 4th or early 5th centuries AD, while the latter shows the presence of north Germans at this late stage (perhaps estate workers, *foederates* or those who occupied the site in the immediate aftermath of the departure of the Roman occupants). That the villa did quickly fall out of use, and then from local knowledge, is testified by the fact that none of the modern local field systems bear any relation to the site of the villa, with one substantial tree-lined field boundary actually bisecting the main range of the second phase.

Figure 35: Tin alloy disc brooch found at Teston villa site. Simon Elliott.

Meanwhile occupation period brickwork has been found in the Medway riverbank immediately downriver of the villa site by the author during a walkover survey (at TQ 7028 5296, see Figure 34), potentially indicating a brick wharf of a type similar to that found at Richborough (Peacock, 1977, 245).

The large scale of this site is indicated by the fact that for a substantial distance either way along the A26 Tonbridge Road (to the north of the villa) neighbours report finding extensive Roman material when gardening (including roof tile, box flue tile and coins, examined by the author and KAFS), this being confirmed by

Figure 36: Hypocaust channel, bath house adjacent to main range, Teston villa site. Simon Elliott.

archaeological evaluation and test pitting which also found an occupation-period flagstone floor surface in one garden (Elliott, 2013, 44). Additionally, a full field walk led by the author in 2015 in the field to the immediate east of the villa site after ploughing (the first time in recent memory) found substantial amounts broken roof tile in the top soil together with pieces of Roman glass and painted wall plaster. Meanwhile, when the adjacent waste-water treatment works to the west of the villa site was built in 1965, a cremation burial and a dressed ragstone wall of Roman provenance were also found, the latter aligning with the flag-floored building found in one of the garden test pits.

Overall, all of the above data indicates a site which, at least in certain phases of its occupation, was comparable in size to the villa estate at Eccles and perhaps even larger.

Wateringbury/ Mereworth/ Nettlestead/ Yalding/ Hale Street/ East Peckham/ Golden Green

Date: Undated.

Type: Unkown.

Site History and Economic Evidence: Moving sequentially upriver from Teston and into the Weald, there is little well recorded evidence of activity during the Roman occupation. Wateringbury has a single *Aureus* of Tetricus listed in the Kent HER as being found to the north of the village in the 1970s (at TQ 6973 5506). Meanwhile nearby Mereworth, on the high ground above Wateringbury and Nettlestead, has occupation period pottery, tile and coins listed in the Kent HER at TQ 6650 5323, though with little detail excepting some of the pottery was Samian ware (Kemble, 1856, 403). These finds were associated with a barrow burial though evidence of the latter (apart from some remaining stonework) was lost when the palladian Mereworth Castle was constructed.

Continuing upriver, there is no evidence at all of an occupation-period presence at Nettlestead or Nettlestead Green (a walkover survey and a search for re-used building material having been carried out by the author, together with extensive desk research), despite their prime locations on the Medway. The same story of a Roman absence applies to Yalding, while Hale Street has a single 3rd century AD Roman coin listed in the Kent HER as being found in a 2004 walkover survey by the Kent Archaeological Metal Detecting Support Unit. Next we get to East Peckham which has a long association with the finding of a two handled Roman vase in the 19th century (reported in *Archaeologia Cantiana* by Payne, 1893, lviii). This was dredged from the Medway between its confluence with the Bourne at East Peckham and nearby Golden Green and marks the furthest occupation period find upriver on the Medway (and at TQ 63 48, though see reference to Great Cansiron in Section 3.1.4).

It should finally be noted here that while Hadlow and Plaxtol on Medway tributary the River Bourne feature occupation-period activity, with three villas and an Ightam Stone/ Gault Clay quarry at the latter, they are not included in this list of primary evidence sites given their economic alignment to the north with the nearby Darent Valley rather than south towards the River Medway.

East Malling Villa (at TQ 7033 5696)

Date: 1st through 4th centuries AD (based on pottery and coin data).

Type: Villa settlement.

Site History and Economic Evidence: This villa site was first located to the immediate east of St James the Great church in 1955 during trial excavations, it then being the subject of a limited investigation by Maidstone Museum (Pirie, 1957, 228). It was next fully excavated in 1965 by Wacher (1965, 257).

Considered here out of sequence given it is located away from the Medway Valley, it is relevant to this research as it occupies a location of strategic importance equidistant between the fords at Aylesford and Barming, sitting on the high ground above the Maidstone bend of the river. In terms of chronological detail, while evidence was found of LIA agricultural activity during the earlier excavation, this went unrecorded at the time and was first noted by Wacher (1965, 257). Three phases of Roman occupation followed in what appears to have been a fine Roman villa featuring painted wall plaster, glass and mosaic tesserae, the site falling out of use in the 4th century (Wacher, 1965, 257).

4.2 The Medway Valley Ragstone Quarrying Industry

The *metalla* ragstone quarries of the upper Medway Valley provided much of the building material for the urbanisation and early fortification of the south and east of occupied Britain (Pearson, 2002, 82a). In the opinion of the author they represent one of the best examples of industrial supply and demand in pre-modern Europe, with a vast industry rising quite literally out of the ground from nothing to cater for those building the new urban environments of Roman Britain, particularly London. In this section I consider this industry in detail, it being a key part of the wider research given its importance to the exploitation of regional natural resources during the occupation. I will specifically reflect on the geology of the stone extracted, discuss why it was so popular as a building material, examine the scale and distribution of its use, and detail the chronology of the quarrying activity. The location of each individual quarry is discussed separately in Section 4.3 where each candidate site is considered in turn to determine its Roman provenance, while in Section 4.4 I consider who the associated workers in the upper Medway Valley *metalla*. Meanwhile, the important question about the role of the state and the *Classis Britannica* with regard to this industry, a key feature of the regional Imperial economy (see Section 2.2.2), features prominently in the discussions of Chapter 6.

Kentish ragstone sits within the Hythe Beds of the Lower Greensand formation. This grey-green limestone outcrops in three regions in the South East, principally in the region of Maidstone (where the best quality stone is found), near Sevenoaks, and at the eastern extremity of the Greensand Ridge where it outcrops in the cliffs near Folkestone. The vast majority of the ragstone used in the Roman built environment originated from the Maidstone outcrop, this being proved through the inspection of numerous examples from across Roman London and the South East (Watson, 2004, 267). Such examination was carried out through petrological analysis using data showing that ragstone from the Maidstone region uniquely features *Trigoniids*-type fossils specific only to this outcrop (Worssam and Tatton-Brown, 1993, 101).

The next question to be addressed is why ragstone was used so extensively in the occupation-period as a building material as opposed to brick or other alternatives. The principle reason is that it would have been easier to build in the stone given its local availability and the fact that vast quantities of brick and tile were already being manufactured from locally accessible materials, stretching the availability of the latter to the limit (especially early in the occupation when unfired clay was also being used in the construction of flooring and unfired walls, Perring, 1991, 63). Pearson (2006, 70) adds that the key factors determining the type of materials used for building during the occupation were economic considerations (for example availability and transportability), suitability (ease of working, durability and similar) and aesthetics, with Medway ragstone meeting all of these criteria as far as the south and east of the province was concerned.

The sheer scale of the ragstone *metalla* in the upper Medway Valley is particularly striking, with enormous quantities of stone being quarried and then transported to its place of use (this providing excellent evidence for the sophistication of regional market integration). The best example of the output of this intensive industry remains the original late 2nd/ early 3rd century AD 3.2km land wall circuit of London. Merrifield (1965, 48) estimated the ragstone volume for this original circuit to be around 35,000m^3, while Hall and Merrifield (1986, 28) later said the wall would have comprised over one million squared and dressed ragstone blocks with an inner and an outer facing, together with a rubble ragstone core which was then set with mortar. Based on the 100,000 man days estimated to have been needed to build the much shorter 760m wall circuit at the Saxon Shore fort at Pevensey (Pearson, 1999, 102), the author estimates some 420,000 man days would have been required for the Roman land walls of London.

Marsden (1994, 84) also believes that vast amounts of ragstone from the Medway quarries were used in London for the construction of the basilica, *forum*, at least three public baths (located at Huggin Hill, another at Billingsgate near today's Lower Thames Street and another on Cheapside), the building suggested to be the Governor's palatial residence beneath modern-day Cannon Street, and many other public buildings in London. Bateman (2011, 31) adds the second phase of the amphitheatre in the north west corner of city, saying: *"Rebuilding the main elements of the amphitheatre included the walls around the arena and flanking the entranceways. The new walls were constructed of Kentish ragstone."*

The monumental scale of all of this work and the planning required in advance to supply the vast quantities of ragstone cannot be underestimated. The second phase of London's basilica and *forum* alone, which date to the late 1st century AD (Hall and Merrifield, 1986, 10), were the largest single stone built structure north of the Alps during the occupation. Marsden (1994, 82) believes this level of societal commitment indicates the involvement of the Procurator in this building programme, including the supply of raw materials (see discussion on the role of the state and the *Classis Britannica* in Chapter 6). It should also be remembered at this point that the ragstone was not just being used for such grand projects but also to facilitate the more day-to-day aspects of the built environment. Examples include the 2nd century AD town house located at Lower Thames Street which preceeded the bath house there (see above) where ragstone was used as the principal foundation and building material (Rowsome, 1996, 421). Meanwhile, ragstone cobbles were also used as the foundation of the walled mausoleum associated with the discovery of a large oolitic limestone-carved eagle found in 2013 in the Minories outside the Roman Wall to the east (Pitts, 2014, 9). Most recently, two post-Hadrianic fire ragstone-built fine town houses have been found on the eastern banks of the Walbrook by MOLA (Watson, 2015) as part of the Bloomberg excavation.

Upper Medway Valley ragstone can be found across the region, its use as a building material not just restricted to London. Examples include the occupation-period walls of Canterbury (see below), the stone-built walls of Rochester

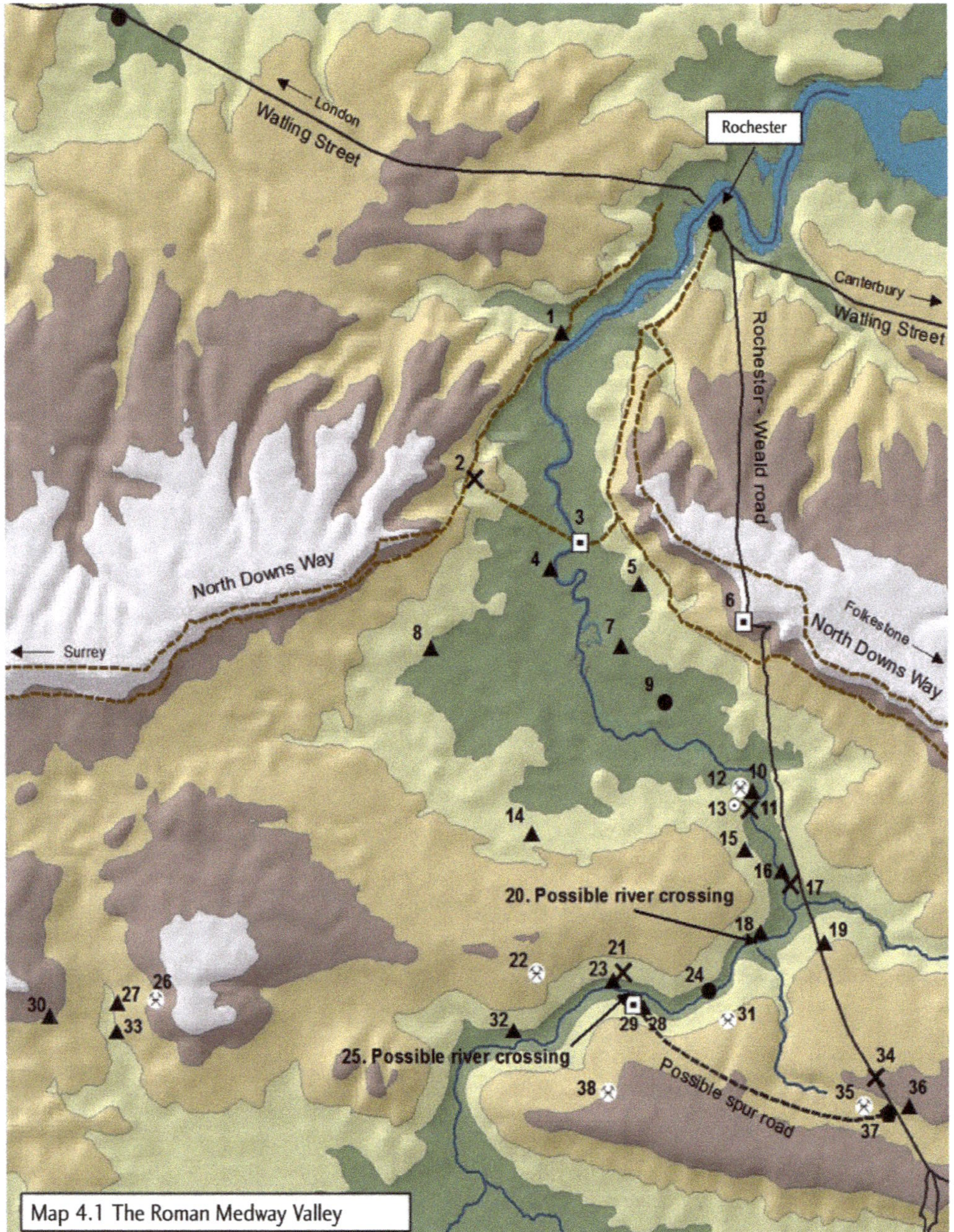

Map 4.1 The Roman Medway Valley

KEY	SITE	KEY	SITE	KEY	SITE
1	Cuxton	14	East Malling	27	Plaxtol
2	Halling	15	Buckland Lane	28	East Farleigh
3	Wouldham/Burham	16	The Mount	29	East Farleigh
4	Snodland Villa	17	S. of the Mount	30	Plaxtol
5	Court Road	18	Florence Road	31	Dean Street
6	Blue Bell Hill	19	Barton Road	32	Teston
7	Eccles	20	Tovil	33	Plaxtol
8	Birling	21	Barming	34	Lockham Wood
9	Aylesford	22	Teston	35	Boughton Monchelsea
10	Allington	23	Barming	36	Boughton Monchelsea
11	Allington	24	Medway Stones	37	Boughton Monchelsea
12	Allington	25	Barming	38	Quarry Wood, West Farleigh
13	Allington	26	Plaxtol		

▲ Villa/Possible villa
⬟ Bath house
Temple
Quarry
✕ Burial(s)
Coin hoard
● Other Below 25

Elevation (mAOD)
150
100
50
25
Below 25

Map created by Elizabeth Blanning © Crown Copyright/database right 2016. An Ordnance Survey/EDINA supplied service.

dating to AD 225 (see Section 5.1.4), early Saxon Shore forts such as that at Reculver where the facing of the defences comprised ragstone blocks (Pearson, 2002a, 79), and in many of the villas in the region. This was especially the case in the Medway Valley, examples here including those at Snodland, Eccles, the Mount, East Farleigh and Teston to name a few. Further afield, upper Medway Valley ragstone was also used in the pre-Boudiccan Claudian temple (Houliston, 1999, 163) and 2nd century AD circus (Crummy, 2008, 27) in Colchester and the early Saxon Shore fort at Bradwell (Pearson, 2002a, 82), both in Essex.

In terms of chronology, Jones and Mattingly (1990, 217) believed that state-run quarrying in the upper Medway Valley began early in the occupation, within 20 years of the conquest at the very latest, while Pearson (2002a, 82) goes further in arguing that stone was being quarried here from around AD 50. Evidence supporting such an early start includes the Claudian temple in Colchester detailed above (Houliston, 1999, 163) and the Huggin Hill bath house which dates to AD 70 (Rowsome, 1999, 262).

Greene (1986, 155) argued that at this early date the geology of southern Britain was well enough understood for limestone and sandstone quarrying from a variety of sources to begin, hence such fine quality building stone appearing in Colchester and London shortly after Claudian invasion. Analogously, examples are abundant of early quarrying elsewhere in occupied Britain, with Hayward (2009, 112) pointing out the early quarrying for freestone while earlier Legg (1986, 55) highlighted similar extractive industrial activity in Dorset. Here, high quality Purbeck marble was being extracted to dress regional public buildings and to line public baths (for example in Exeter) shortly after the AD 43 invasion (Bidwell, 1980, 11, and Pearson, 2006, 113). All of these examples support Goldsworthy's (2014, 239) hypothesis of a mercantile presence in the South East of Britain well before the Claudian invasion.

Back to the Medway Valley, the ragstone quarrying quickly developed into industrial scale activity to fulfil the booming demand for building stone as the new stone-built urban environment blossomed, and also to facilitate the construction of the region's new villa estates and roads. This demand peaked in the late 2nd and early 3rd centuries AD when construction began of the region's first fortified wall circuits, the best example here again being the land walls of Roman London. It is the building of this wall that Merrifield *et al* above use to demonstrate the sheer scale of the ragstone quarrying industry that would have been required to meet such demand.

It is also the walls of London however which are our first indication of a dramatic change taking place with regard to the Medway Valley ragstone quarrying industry, in the form of the later river wall built towards the end of the 3rd century AD to close the circuit, together with new bastions around the perimeter. Far from the new walls and towers being built with the carefully worked, uniform ragstone blocks evident in the land wall, they were constructed from roughly reworked local materials, re-used from demolished public buildings and mausoleums. The reuse in the river wall of finely detailed sculptured stones depicting classical scenes is well documented by Blagg (1980, 5), who believed that some originated in a monumental arch featuring ornamental screens. Sheldon (2011, 230) also highlights the reuse in one bastion of material from the mausoleum of the province's third Procurator, Julius Classicianus. Pearson (2006, 30) argues that the reuse of materials rather than building with newly quarried ragstone is a manifestation of the ending of the era of major public building in towns in Britain, excepting some limited activity in York (where at least one large public building was built in the later 3rd century AD) and in Carausian/ Allectan London where another large public building has been interpreted as a possible palace (Williams, 1993, 31) or temple complex (Bradley and Butler, 2008, 9). Even with this latter building however change from the earlier construction regime is evident, it again being built of re-used building material in a similar manner to the river wall and bastions (Rogers, 2011, 96). Similarly, where ragstone was used for the construction of some of the later Saxon Shore forts, it is evident they are actually built from re-used material from earlier nearby structures (with Richborough a prime example, Fields, 2006, 33).

As is evident above, with re-used material taking the place of newly quarried ragstone, a major change seems to have taken place with regard to the Medway Valley quarrying industry from the mid-3rd century AD, with the latest requirement candidate for freshly worked stone being the wall circuit of Canterbury dating to around AD 270. Even here the quantity of ragstone used is limited when compared to earlier regional requirements given it is only visible in the context of footings and lower wall facings (flint being the preferred main facing), with the stone also being potentially re-used. After that time at the very latest, no more would enormous quantities of building stone be extracted in the upper Medway Valley and sent downriver to the Thames Estuary (reasons for this change are discussed in Chapter 5). Pearson (2006, 30) says that this is part of a broader picture of quarrying and building in the later provinces and *diocese*, where such activity by that time is small in scale and one of repair and refurbishment when compared to the intense urbanisation of the earlier period. Everitt (1986, 51) said that in this later, local phase of regional quarrying, one can see the origins of a new local phenomenon in the upper Medway Valley, with the *'great quarries'* of the earlier phase being replaced by much smaller 'stone pits' known locally in more recent times as *'petts'* which catered for local demand.

4.3 Where Were the Ragstone Quarries Located?

Moving on to where the ragstone quarries from the earlier, industrial scale period were actually located through to the mid-3rd century AD, the below review of candidate sites should be seen in the context of them sitting within a densely settled industrial landscape featuring the quarries themselves, supporting infrastructure including iron manufacturing, elite and other settlement, agriculture and an integrated transport infrastructure.

Historically, the association of this region with ragstone quarrying during the occupation has its roots in at least the antiquarian record, if not earlier (Robertson, 1883, 68). The identification of the specific quarry sites has long proved problematic however, with to my knowledge no detailed attempt being made in this regard until my own research was set in train. The main issue has been the longevity of operation of the quarrying sites providing accessible and good quality stone, with for example in a Medway Valley context ragstone continuing to be extracted from some quarries at various stages long after the occupation, thus destroying any direct evidence of Roman activity. This is an issue with Roman quarries across Britain as Russell (2013b, 1) explains: *"In Britain... analysis of built structures clearly indicates where stone was quarried despite the fact that later activity, especially in the Medieval period, has largely eradicated all traces of ancient quarrying."*

Such a problem with the Medway Valley quarries is detailed by Goacher (2012, 12), while Pearson (2006, 45) also highlights the issue of the modern re-use of quarries for landfill, this certainly being the case for the section of the Dean Street quarry detailed below closest to the River Medway. Other regional quarries, particularly that at Boughton Monchelsea, have had data from occupation-period quarrying activity obscured by later settlement in the quarry bottoms, though the flooding of quarries for recreational use or to serve as reservoirs does not appear to be an issue in a Medway Valley context.

Despite such issues I have persevered in my search for the quarries using many kinds of data, in the first instance starting with a review of the 1797 OS maps of Boxley, Maidstone and Addington to identify industrial scale quarries which by that time were clearly extinct and overgrown (and in the case of Boughton Monchelsea with settlement already beginning to appear within its base). With the candidate quarries thus identified, I then developed a methodology to test their likely Roman provenance. This comprised the following:

- Appropriate quarry forms.
- Proximity to villa and non-villa settlements and other industrial sites.
- Proximity to occupation-period transport infrastructure, including roads and riverine structures.
- Proximity to other sites of archaeological activity including cemeteries, burials and coin hoards.
- Analogy from elsewhere in the Empire.

Specific comment should be made here in particular with regard to the correlation between the candidate quarries and the villas of the upper Medway Valley. In her definitive review of rural settlement in occupation-period Kent, Blanning (2013, 195) says of these villas that there is little relationship between these sites and the most fertile agricultural soils in the valley. Further, while some are associated with a change in the bedrock which would have promoted agricultural diversity (as with the villas along the Holmesdale elsewhere in the county, for example at Thurnham, Foreman, 1999, 3), others are not. These include the villas at Allington and East Farleigh, while those at Florence Road/ Bower Lane in Maidstone and at Barming are only marginally linked to a change in bedrock. Therefore something else is in play here, which I argue below is an association with the extractive industries exploiting natural resources.

Based on the above wider forms of data, and with the additional support of personal communications from local people with expert knowledge in various fields, I now list each candidate quarry in turn as one heads upriver. At Section 4.3.6 I then consider the provenance of each. To provide context when deliberating the sites, analogously each of these quarries (if proved to be Roman) would have been of the scale of Peacock's manufactory means of production of pottery in the Roman world (1982, 10), also having parallels with his military and official modes of production.

4.3.1 Allington Quarry

This quarry is located actually on the modern tidal reach at TQ 7446 5792, to the immediate west of Allington Castle. Its area based on the 1797 OS map of Boxley, at which time it was extinct and overgrown, was 61,600m^2.

Allington is often cited as one of the main sources of quarried building ragstone during the occupation, though without the exact quarry site being identified. Supporting data suggests a significant Roman presence including the location in the immediate vicinity of a Roman villa settlement, coin hoard and burials, all detailed fully above in Section 4.1.4. A local tradition also talks of the finding of a Roman 'village' associated with the ragstone quarrying, though there is no data to support this in either the antiquarian record or the HER. Anecdotally however, one can reasonably speculate that as this is the most accessible site to extract ragstone in the Medway Valley (it outcropping naturally and being on the tidal reach), it would have been one of the first exploited during the occupation.

4.3.2 Boughton Quarries

The next major site for consideration is Boughton Quarries in Boughton Monchelsea, above the tidal reach. Again extinct and overgrown on the 1797 OS map of Maidstone (by which time settlement is already evident in its base), this quarry has an area of 54,600m^2. Reflecting the fine quality ragstone outcropping here along the valley of the Loose Stream, quarrying was restarted in the area in the 19th century with the Brishing, Middle and Furfield Quarries being opened to the east of the original extinct quarry and Beresford Quarry to the west, stone extraction continuing here until the 1930s. Boughton Quarries itself was not reopened however and the later quarries were evidently much lesser in scale.

Physical evidence of occupation-period quarrying in Boughton Quarries is provided by the vast sheer quarry faces and extraction terraces evident in the pre-modern workings (some 300m in length), already overgrown and settled as detailed above by the later 18th century. Further, as set out in Section 4.1.4 above, there are many nearby sites of Roman provenance, including villa and non-villa settlement, the famous Lockham Wood walled cemetery and the equally famous remote bath house. Meanwhile, in terms of occupation-period transport routes, the Wealden road from Rochester to the Beauport Park iron-working site actually bisects the string of ancient and modern quarries here, with its modern manifestation in the A229 being called here the Loose Road.

Traditionally, the wider region which runs up to and over the Coxheath into the Weald is known as the Quarry Hills (this pre-dating the opening of the later quarries, Hastings, 2000, 9), and there is much local association with ragstone quarrying during the occupation. An example is to be found in the official history of St Peter's church in Boughton Monchelsea which speculates that it is located on the site of a Roman temple (though with no provenance), while Hastings (2000, 9) says in the official Parish history that:

"In the absence of agrarian expansion, most wealth in Boughton in Romano-British times seems to have come from exploitation of the ragstone quarries running from Brishing towards Loose...most Romano-British remains in Boughton are presumed to be connected with the quarrying of ragstone."

4.3.3 Dean Street Quarry

The next quarry as we move up the river valley is that at Dean Street, the roadway which links Heath Road in Coxheath (the high ground separating the Medway Valley with the Weald) with Tovil Road on the outskirts of Maidstone (and adjacent to the Medway). Again above the tidal reach of the River Medway, the quarry is centred at TQ 7450 5334 and runs parallel to the road, beginning just below Coxheath and then running for an impressive 2.6km down to the Medway by which time the gradient is around 1.40. It has a striking quarry form, being long and narrow (with a maximum width of 166m wide at its head, see Figure 39, though widening out to 255m as it approaches the river), and given its length has a huge area, this being 356,400m^2.

This enigmatic site, once again extinct and overgrown on the 1797 OS map of Maidstone, already had local notoriety by the time Coles (1839, 126, see Section 5.5.3) was writing his early 19th century guide to Maidstone, he commenting in a bespoke section on its length and steep sides and noting its puzzling lack of water. In the 20th century the quarry was famous in geological circles for the then readily accessible faces of ragstone of varying quality which were visible along its length down to a depth of 37m (Daniels, 2015, 6), though in recent years this has been limited to just one face adjacent to Passmore Way in Tovil given the reuse of much of the site nearest the river as the Maidstone municipal rubbish dump.

Known locally as 'The Roman Quarry', the site was brought to the attention of the author by a local farmer who was convinced of its provenance as a Roman site. Data to support its the occupation-period provenance comes in a variety of forms, for example the long quarry profile with its impressive sheer faces which have clearly been worked on an industrial scale. In fact scale is the first determinant hinting at its Roman origins. Darvill and McWhirr (1982, 137) point out that while quarrying was in evidence in Britain before the occupation, the magnitude of operations increased dramatically after the arrival of the Romans. In this context the Dean Street site would be the ultimate example given its huge size. It is certainly much bigger than previous claimants to being the largest occupation period quarry in Britain such as the 'huge' 0.75km long Corallian Limestone quarry located in 1988 at Blunsdon Ridge near the occupation-period site at Abbey Meads in northern Wiltshire (Walters, 1988, 8). The Dean Street quarry's potential to be of great significance to the wider study of industry in the Roman Empire is also evident in Mattingly's (2011, 170) comment that the main opencast gold mine at the Las Medulas mining complex in north western Spain was, at 2.5km across, one of the biggest preindustrial man-made hole on the planet. The Dean Street site is bigger (at least in length), and in terms of human endeavour it can perhaps be spoken of in the same terms as Hadrian's Wall (where by way of comparison the average quarry, such as that at Combcrag Wood in Cumbria, was less than 0.5kms long), the Antonine Wall, the Foss Dyke and Car Dyke as a monumental example of Roman engineering and industrial expertise and enterprise.

For hard data supporting the Dean Street quarry's Roman origins there are a number of immediately adjacent occupation period sites, for example the *'Roman foundations and coins'* detailed in the Kent HER (number TQ 75 SW 251) adjacent to Harps Wood to the immediate east of the Dean Street quarry. Found in 1860, this site is now visible as a crop mark on the May 2015 Google Earth pass (at TQ 7481 5242) and also on the recently released EA LIDAR imagery of the Medway Valley (see Figure 40 below) and appears to show a multiple room building of similar design to a Roman barracks. An investigation is planned in the near future. Meanwhile, to the immediate west of the quarry at Pimpes Court (TQ 7535 5275, detailed under the Boughton Monchelsea 'Remote Bath House' entry in the site list at Section 4.1.4) Samian ware, cinerary urns and other material culture finds have also been located (Taylor, 1932, 113). Most recently the Gallants Lane potential occupation-period iron-working site has been located to the west of the quarry, this detailed above in Section 4.1.4.

Figure 37: Milestone/ grave marker, found along the length of the Roman road linking the Dean Street quarry with the Barming ford/ East Farleigh villa. Simon Elliott.

Figure 38: Rim of Gauloise 4 amphora from burial group along Dean Street Quarry road. Simon Elliott.

Meanwhile a further important piece of data in terms of physical evidence comes from the author's location of the Roman road detailed above in Section 3.5 which, at 3km long, specifically links the quarry to the East Farleigh villa and Barming ford (Elliott, 2013, 40). Clearly visible (at TQ 5735 5219) leading towards the Medway from the Horseshoes pub and stables situated on Dean Street, the road has an *agger* partly covered by a substantial and ancient hedge line (which is incorporated into Fright Wood on the 1797 Addington OS map) and runs across Gallants Lane (the site of sepulchral burials, see 4.1.4 above) into Kettle Lane, from where it descends along a shallow slope to the river. Along this length of road the author has additionally located a milestone or grave marker (see Figure 37) and evidence of further occupation-period burials (see Figure 38).

Next, research is now being carried out into another site which would link the quarry to another elite settlement. This is the potential occupation-period bridge crossing at Tovil (Elliott, 2013, 40, and see Tovil entry in 4.1.4 above) which would link the Dean Street quarry at the point it meets the river to a roadway on the opposite bank leading to the Florence Road/ Bower Lane villa settlement in Maidstone. In this context a strong case can be made that either, or both, the East Farleigh villa (featuring the strongest evidence) and the Florence Road/ Bower Lane villa were home to the elites who ran the Dean Street quarry (especially given the weak agricultural links to the location of these villas, Blanning, 2013, 195, see Section 4.3).

Most recently however, a key piece of physical evidence providing hard data for occupation period quarrying at the Dean Street site has been discovered by the author in the form of Roman stonemason tool marks on the only remaining section of exposed antique quarry face. This is next to the residential car park of Rockwell Court off

Figure 39: Section of the Dean Street quarry today, mid-length, re-used as a local fine garden. Simon Elliott.

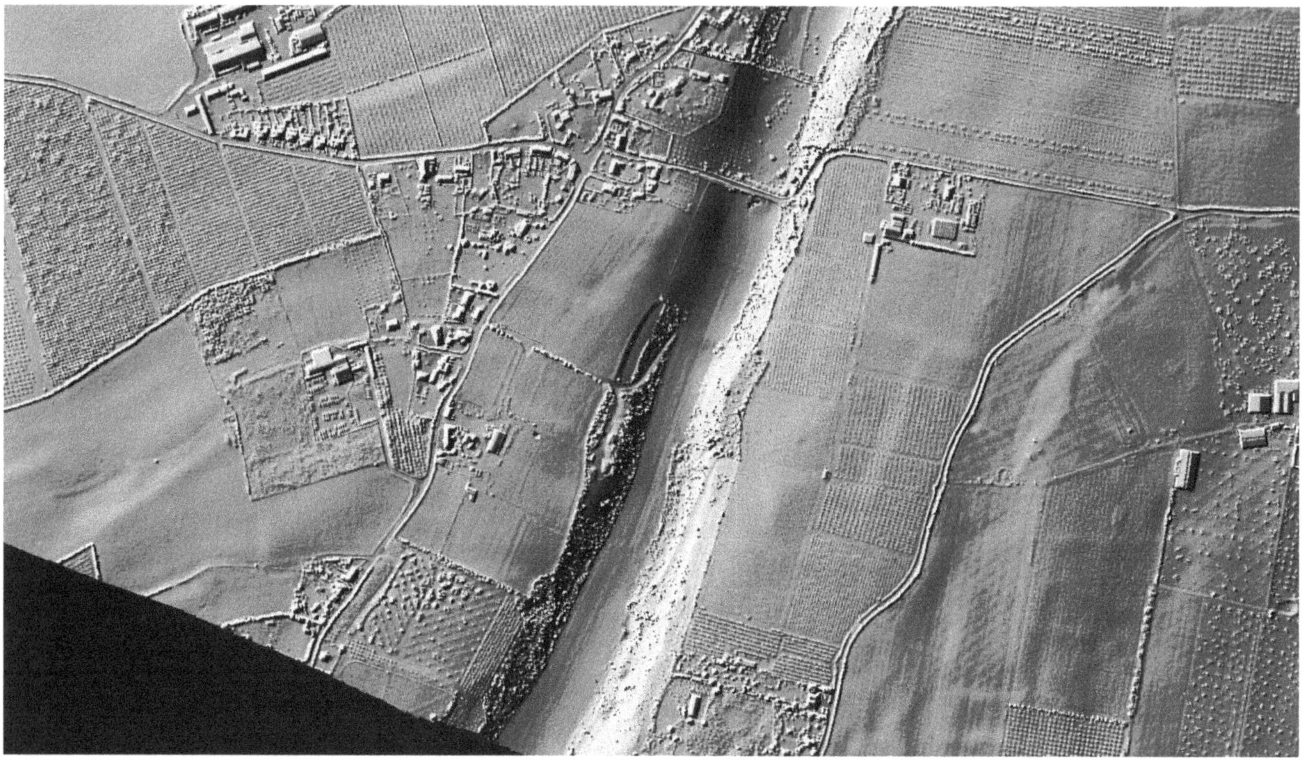

Figure 40: LIDAR image of the Dean Street quarry mid-length, showing the huge impression it makes on the local landscape. The potential multi-room building is at bottom right. Supplied by Dr Paul Wilkinson, public domain image, Environment Agency.

Passmore Way in Tovil (at TQ 7523 5420). Here can be found regular chevron-shaped markings which have been identified by the author as Roman, this published here for the first time (see Figure 43).

Further research has suggested they were made by masons using either a bull-nosed or flat-headed chisel (Blagg, 1976, 158). Such occupation-period mason's marks are unique in Kent, though by way of analogy such chevrons appear on quarried Roman stone blocks examined by the author at Roman sandstone and limestone quarry sites in Paphos in Cyprus (see Figure 44).

Meanwhile, in an earlier bid to find supporting data at the Dean Street site, three auger holes using a manual 2m augur were drilled into the valley floor by the author in 2009, one at each side and one in the centre (the latter at TQ 7433 5220), with all being 100m from the head of the quarry. Those at the side revealed very heavy clay loam to a depth of 1m before hitting solid rock. That in the centre had the same composition down to 2m, again then hitting solid rock. To ensure that this latter result was not a unique event two further holes were drilled, 5m north and south off centre, with both featuring the same result. These findings match those of the more recent 2012 auguring exercise of Spencer (2013, 47) further down the quarry (at TQ 7474 5320), also using a manual 2m augur, he also reporting the concaved nature of the valley floor. The author believes that the auguring findings are indeed evidence that it was an occupation-period quarry, given the extensive quantities of alluvial hill wash which have accumulated in the valley floor. Additional confirmation of the presence of this deep layer of alluvium comes from Ward's (1999, 1) report for CAT on the 1999 archaeological watching brief at Tovil Mill where the Dean Street quarry descends towards the Medway, he reporting that within a 1.7m deep observation trench: *"...brown clayey loam was...visible. This latter soil was probably not made ground and may consequently be the result of hill wash..."*

Finally, additional support for the interpretation of the Dean Street site as a Roman quarry could also be provided if the 'Medway Stones' detailed in 4.1.4 are ultimately proved to mark the location of an occupation-period wreck carrying worked stone, the Dean Street quarry meeting the river almost at the point where the stones were found.

While there is little doubt that the Dean Street site featured major quarrying activity, there is some debate about whether it was purely man made or whether the site was an unusual natural feature which was exploited from the occupation onwards for its readily available ragstone. To facilitate this debate the author helped organise a visit to the quarry by a large gathering of geologists and archaeologists in May 2010, with the resulting opinion being divided between the man made and the natural-but-exploited interpretations.

Most recently, in an in-depth study of the site to determine its Roman provenance, Spencer (2013, 55) has concluded that it is actually a rare post-glacial natural feature which has been extensively exploited for ragstone quarrying from the occupation onwards (this explaining the unusual quarry form). He says:

"The gentle dish-shape of the rock surface of the valley bottom, together with the contour pattern at the head of the valley suggests that the original valley feature was almost certainly carved out by a periglacial stream formed during the last Ice Age, subsequently becoming a dry valley once the permafrost conditions thawed."

He adds that another option for the natural creation of the quarry would be cambering whereby a heavier layer of later exploited ragstone sat above an original plastic layer of clay which was squeezed into the Medway valley below.

However the site was created though, Spencer (2013, 55) is clear that it was then heavily quarried, he saying: *"There is extensive evidence of quarrying along the entire valley which has widened and steepened the valley sides."*

He has identified three distinct periods of activity in the quarry. These are:

- A modern phase in the 1930s (with a focus on the lower part of the valley near the river) which was industrial in scale and removed entire seams of ragstone.
- A 19th and early 20th century phase which was much more localized in scale.
- Finally, a pre-18th century phase when once again extensive industrial scale quarrying is evident.

Spencer believes (2012, 55) that the monumental scale of the quarrying carried out in the pre-18th century and earliest phase was only possible in the occupation period as it is too vast in scale and enterprise to be medieval. As evidence of this scale he says that for each 1m thickness of stone removed from a given seam, some 60,000 tonnes of ragstone and associated hassock would have been produced.

The author is in agreement with this latter interpretation, with the Dean Street quarry being by some way the most important in the sequence of ragstone quarries in the upper Medway Valley during to the occupation.

4.3.4 Quarry Wood, West Farleigh

The next major quarry in the valley sequence is that at Quarry Wood where Heath Road hits the head of the Ewell Valley in West Farleigh (the quarry located at TQ 7194 5193). This enormous antique quarry, once more above the tidal reach, was already extinct by the time the 1797 Addington OS map was created, its base covered in ancient woodland. With an area of 215,000m^2 it is the second largest in the sequence of upper Medway Valley quarries after the Dean Street quarry and is called Quarry Wood on the OS maps from the 1870s sequence onwards, though it remained extinct.

Once again scale and the impressive quarrying terraces (some over 500m in length) indicate an industrial scale of operations in the pre-modern era, while in terms of hard data the site sits within the same landscape of Roman occupation as the Dean Street quarry, it being located just above the Roman road connecting the Dean Street site to the East Farleigh villa and Barming ford. The Gallants Lane potential iron working site detailed in Section 4.1.4 is also close by, with the nearest likely cinder field only 200m away.

Ragstone quarried here would have been carried the short distance down the Ewell Valley to the Medway for onward shipment. Today the quarry is a nature reserve managed by the Kent Wildlife Trust and is listed as a Regionally Important Geological Site (RIGS) given that exposed ragstone cambering is still visible in the quarrying faces.

4.3.5 Teston Quarry

Moving upriver again, the next and final quarry reached is that at Teston (at TQ 7045 5425) which, by the time of the 1797 Addington OS map, was fully extinct with a trackway meandering northwards up its length towards Barming Heath and with fish ponds dug into the base of its southern extremity to supply nearby Barham Court country house. Again above the tidal reach, it has an area of 35,830m^2.

The quarry's occupation period origins are identifiable once again by the scale of the sheer faces and extraction terraces (some 700m in length) in the pre-modern workings, and analogously due to its close resemblance in form to the Dean Street quarry with which it shares a similar geology and hence shape (see Figure 41 for cross section). It also shares with the latter the close association between the quarry and a nearby villa, this time the very grand 1st through early 5th century AD example at Teston less than 1km away and which is detailed in Section 4.1.4 (Elliott, 2013, 40). Additionally, between the quarry and villa here there are a number of occupation-period burials detailed in the HER. These include that at TQ 70 54 found in 1973, featuring a mid-2nd century AD cremation burial in a Patch Grove Ware jar which included Samian ware

Figure 41: The Teston Roman ragstone quarry today, viewed looking towards the head of the quarry. Simon Elliott.

pottery. This burial site is 500m away from both quarry and villa settlement as one heads down the slope of the Medway Valley from the former to the latter.

4.3.6 Short Consideration of the Evidence

The starting point for this consideration is that, as set out above in Section 4.2, we factually know that the ragstone used in the occupation to help construct the built environment in Roman London and elsewhere in the South East definitely outcropped in the upper Medway Valley. The question is therefore not whether it was sourced from here, but specifically where from? In that regard, and based on the data set out in Section 4.3.1 through Section 4.3.5, I believe that a very strong case can be made that the Boughton Quarries, Dean Street quarry and Quarry Wood have Roman provenance, this especially the case given they were all set within what was clearly a densely settled occupation-period landscape featuring elite and other settlement, other industrial sites and with transport infrastructure evident both in terms of roads and the River Medway (illustrating the regional level of market integration). Similarly I believe that Allington Quarry and Teston Quarry were also likely Roman quarries, though I acknowledge that the data in both the latter cases is slightly less in quantity and nature.

A final point to note here with regard to all five quarries is that during the occupation all would have been above the tidal reach of the river (see Section 4.1.3, especially with regard to Allington Quarry which is on the modern tidal reach), where without riverine infrastructure (as I argue existed during the Roman period in my 'Medway Formula' research, see Section 4.5.2) the Medway would have been unsuitable to support such large scale industrial activity and associated settlement. Indeed, in the modern era such riverine infrastructure did not exist until the later 18th century at the very earliest and was associated with the opening up of navigation on the upper Medway by the Medway Navigation Company. I therefore believe that given the scale of the quarries discussed here, this fact further adds to the data set out above locating them chronologically to the Roman occupation.

4.4 Who Were the Workers in the Medway Valley Ragstone Quarrying Industry?

Given the close parallels in an industrial context between the *metalla* in the Medway Valley and that in the eastern/coastal region of the Weald, one would expect that the workforce structure would also have been similar, with organisation along comparable lines (note the different settlement patterns between the two regions and the impact of this on interpreting the available data in the context of their role in the Imperial and provincial economies is discussed in full below in Section 5.1.3 and 5.1.5).

At the top once again would be a *procurator metallorum*, perhaps the same individual in charge of both *metalla* as discussed above in Section 3.4. Directly below him would again be his staff tasked with the actual management of the quarrying and all associated activities, including more junior *procuratores*, equestrians, freedmen and the military personnel seconded by the Governor. These middle-ranking officials fulfilled a variety of roles, for example the *philosophus* in charge of transport, the *officinator* tasked with running a specific individual quarry, and within each quarry the *probator* in charge of selecting the best blocks (Pearson, 2006, 40).

Below these senior officials, once again there were a range of options available to directly manage the *metalla* operations, ranging from the state chain of command continuing downwards to carry out the work themselves (possibly in this case in the form of the *Classis Britannica,* see discussion in Chapter 5), through to all the work being outsourced to civilian contractors (though again all would have reported upwards to the *procurator metallorum*). Finally, looking to those who carried out the physical 'coal face' work itself (whether skilled or unskilled), we again find semi-autonomous individuals, groups of freedmen,

Figure 42: Roman quarrying tools in action. Saw marks on sandstone blocks in Roman quarry, Paphos, Cyprus. Simon Elliott.

Figure 43: Close up detail, marks made by Roman flat-headed chisels, exposed section of Dean Street quarry. Roman *metalla* in action. See section on workforce below. Simon Elliott.

Figure 44: Roman quarry face with chevron-shaped stonemason tool marks evident, sandstone quarry in Paphos in Cyprus. Simon Elliott.

indentured workers from the local population operating under imposed *munera* and forced labour (criminals, captives and slaves).

Within this broader picture of the workforce in the occupation-period Medway Valley ragstone quarrying industry, we are fortunate that there exist numerous analogies from state-run quarrying enterprises across the Empire that help paint an even more detailed picture of how these administrators and workers would have experienced their lives at all levels.

Starting with the administrators below the level of the *procurator metallorum* and his team, there are many examples of the state carrying out the activity itself through the use of the military, and similarly of the state subcontracting the work to private contractors. In the case of the former, Hirt (2010, 329) says that archaeological data from the huge Roman marble quarries at Simitthus in North Africa shows no sign at all of any sub-contracting, with the state officials having firm control of operations. By way of analogy, the scale of the upper Medway ragstone quarrying industry (and to a lesser extent the iron industry in the coastal/ eastern region of the Weald) seems to be a directly comparable case here (this being factored into the discussion regarding the role of the *Classis Britannica* in Chapter 5).

In the case of the latter, Russell (2013a, 47) provides a useful analogy in the marble quarries at Dokimeoin and Teos in Asia Minor and at Chemtou in Tunisia. Here, unused blocks have been found with the inscriptions of individual *rationarii* (account holders) who would have been operating under *locatio conductio operis* contracts to meet contractual obligations to the state. Meanwhile, using the exploitation of metal resources as another analogy, we can also refer to Alba Iulia in modern Romania where epigraphic evidence on monuments and wax tablets indicates that the state engaged private quarrying companies to work on its behalf at the large gold mining complex there, a site which attracted specialists from Illyria to Dacia (Hodges, 2014, 55). Greene's (1986, 146) review of Roman gold mining in Spain also details a consensus of academic thinking in favour of sub-letting in that geography to smaller leaseholders, though to my mind the massive scale of operations at Rio Tinto may have been too great for such a complex sub-letting exercise, and might have needed direct control by the state.

Whether the administrative work was carried out directly by the state or outsourced, from a phenomenological perspective Wadi Umm Wikala in Egypt provides another useful analogy. Spencer (2013, 17) explains that here operations were run from a *principia* building similar to that found in many Roman fortresses which would have managed the on-site *officina* workshops/ studios detailed below. Other buildings at Wadi Umm Wikala included a shrine or temple, a building featuring a cistern and a *hydreuma* fortified well. We can add to this in a Kentish quarrying context a remote bathhouse such as that argued for Boughton Monchelsea (see Section 4.1.4).

Moving below the administrative level to the workforce, Greene (1986, 148) said that skilled workers would have been attracted to the quarrying enterprises because of the profits to be made, especially at the larger state-run sites. There were lifestyle perks to be had also, with insight here provided by the Vipasca tablets which detail the contractual aspects of the leasing arrangements in the Spanish mining industry. Greene (1986, 148) said in this regard that the commercial emphasis of such arrangements gives the impression of striving to offer essential amenities to the skilled workers as a means of making such a commercial undertaking attractive, including bathing facilities.

Pearson (2006, 67) explains that expert masons would have been much in demand, with many able to afford funerary memorials or sarcophagi. An example would be the stone worker Priscus who had an altar dedicated to his memory in Bath and who, intriguingly, appears to have originated in Chartres in modern France, this illustrating the distance skilled craftsmen would travel for employment (Pearson, 2006, 67). Another analogy would be the mining industry in Galicia where inscriptions indicate workers originated from across the peninsula (Mattingly, 2011, 170). Skill was certainly required by part of the workforce, at least to work the quarried stone before transportation, with both Greene (1986, 152) and Blagg (2002, 8) saying that considerable preparation was carried out at the quarry site itself, if only to save weight. This is certainly indicated by the 'Medway Stones' detailed in 5.1.4 above which, as blanks, were part finished. Similar insight into the agency behind this skilled activity is provided by the ragstone coffins found at the Roman cemetery at Keston near Orpington. Mynott (1978, 1) detailed their fine quality, with the specimen found in 1938 having a separate coffin and flat lid, each carved from a single large piece of ragstone. She speculates that, given the quality of the stone, an origin in the Maidstone area is likely and says that such goods would have been made to order as rough outs which were finished at their place of use. In this case, given an Upper Medway origin, the 0.5 tonne blanks would have been transported down the Medway, up the Thames Estuary and then up the River Cray before a short over-land journey to Keston.

Russell (2013a, 47) says that there was a clear division within the quarries between *caesura* workstations where the stone was actually cut and *officina* workstations where the cut stone was then part finished. Within each quarry there would have been multiple examples of both, with the *officinae* able to order up stone from a variety of *caesurae* cutting different qualities of stone. He uses the example of the marble quarries at Dokimeion where inscriptions have been left at some of the *officinae*. Interestingly, the inscriptions also show that the various *officinae* had an association with the specific destination of the stone they were working on. Thus at Dokimeion one has an *officinae smyrnaiorum,* preparing materials for Smyrna, with other *officinae* evidently preparing materials for Ephesos and Nikaea. Meanwhile, Dworakowska (1983, 165) reminds us of the *probator's* overseeing workstation, maintaining quality control in the quarry. She cites as evidence data from worked stone at the Marmorata Imperial marble working yard in Rome, where some blocks have a mark of approval and some a mark of rejection from the *probator*. We can use these analogies to phenomenologically consider the Dean Street quarry in Kent during the occupation through to the mid-3rd century AD, with perhaps an *officinae londinium* ordering and preparing cut stone for use in London and an *officinae cantium* similarly working on stone for use in eastern Kent. In both cases, once signed off by the *probator*, the stone would then be shipped for finishing at its destination.

The skills required by the workmen operating in the *caesura* and *officina* is evident in the range of tools newly available to the workforce compared to pre-occupation period quarrying (see Figure 42). Specific types included (Greene, 1986, 151): *"...a comprehensive range from axes and adzes to fine chisels and files (which) allowed stone to be finished to the precise degree of sophistication that was required for any particular purpose."*

Pearson (2006, 53) says that examples of such occupation period tools and technology have been found across Britain at sites ranging from Vindolanda in the north to Wroxeter in the west and London to the east. Meanwhile, Blagg (2002, 16) highlights the technology available for finer pieces of work, for example the animal and man-powered lathes used when finishing larger pieces of stone for use as column blocks and similar.

Below the skilled workforce one next finds the unskilled workers of the *metalla* striving in their *caesura*, living a similar sad existence to their Wealden counterparts detailed in Section 3.5, perhaps the most unfortunate members of Roman society. The physicality of the work of these indentured workers and forced labour is evident in the fact that quarrying technology remained static until the advent of explosives and mechanical extraction in the 19th century (Ward-Perkins, 1972, 4). We know therefore that quarrying during the occupation would have depended on backbreaking work removing stone blocks using wedges, cutting grooves or drilling holes to encourage splitting (cutting into the rock face from top to bottom so that on a given block only two faces, underside and rear, needed freeing, Ling, 1985, 19). It is once again no surprise that these unfortunates are the least represented in the archaeological record of Roman quarrying from across the Empire.

4.5 Transporting the Quarried Ragstone

As detailed in Section 2.5 above, maritime transport was the preferred means of conveying materials in bulk of all kinds during the Roman occupation and that was certainly the case with regard to the upper Medway Valley ragstone quarrying industry. The fact that all five candidate quarries as set out in Section 4.3 were in close proximity to the River Medway is testament to this, as is the fact that nine out of 10 of the confirmed villas in the wider Medway Valley were located within 1km of the river (Blanning, 195, 2014). To take one specific example, in this case the use of Kentish ragstone in Roman London, Merrifield (1965, 49) was blunt on the subject of the importance of maritime transport, he saying that: *"The best means of transport for bulk of this kind was by boat, and the Medway and the Thames provided a water-way from quarry to City."*

Figure 45: *Codicaria* towed river barge, with forward towing mast. Elizabeth Elliott.

In this section (having considered the river itself above in Section 4.1.3) I look at the maritime distribution of upper Medway Valley ragstone during the occupation, consider whether riverine infrastructure in the form of locks and weirs was utilised to facilitate the exploitation of such natural resources (as I discussed in my 'Medway Formula' research), and finally look at the specific details of a typical journey for a boatload of ragstone to its place of usage during the Roman occupation.

4.5.1 Where Was the Ragstone Transported?

Kentish ragstone was exported widely across the South East during the Roman occupation for use as a building material (and for the construction of roads), as far afield as London, Colchester in Essex and Dover on the east Kent coast (a key indicator of the high levels of regional market integration).

Broadly two types of vessels are candidates for use in transporting the ragstone. The first is a sea-going merchant sailing vessel built in the Romano-Celtic rather than Mediterranean tradition (Milne, 2000, 131), featuring a frame of closely spaced large timbers, flush laid planking, plank seam caulking and a mast step noticeably forward in the hull. Designs of this type have been associated since 1962 with Peter Marsden's investigation of a wreck found next to the most northerly bridge pier of Blackfriars Bridge, this vessel having founded at the confluence of the Rivers Thames and Fleet at a presumed occupation-period wharf in Roman London. Dubbed the Blackfriars 1 ship, this vessel is of particular importance to this research given it was found to be carrying 26 tonnes of Kentish ragstone from the upper Medway Valley when it sank (in a hold capable of carrying up to 50 tonnes, Marsden, 1994, 80). The ship was 14m in length and 6.5m wide, with a shallow draught of 1.5m and a maximum speed of around 7 knots in favourable conditions (Pearson, 2002a, 85). Built of oak, it had no keel but featured two broad keel-planks, a stempost with corresponding sternpost and hazel twig caulking for the carvel planking. The mast was supported by a rectangular socket mast-step in the base of which a bronze coin of Domitian was found. Dendro-analysis has dated the vessel to around AD 140 and identified that it was built in the South East of Britain. A variety of wrecks of this design have been found across north-western Europe of various sizes, illustrating its ubiquity.

The second type of vessel is the *codicaria* towed river barge (Milne, 2000, 131, see Figure 45), even more ubiquitous than the Blackfriars 1 style vessels and found in the archaeological and epigraphic record across the Empire. Built in both the Romano-Celtic and Mediterranean traditions, *codicaria* are heavily referenced by Ausonius (Mosella, 5) who speaks of the use of hawsers attached to a forward-set towing mast being used by men to tow such barges from the riverbank. Selkirk (1983, 83) provided insight here using the River Tiber as an example, detailing teams comprising a helmsman and four *codicarii infra pontem sub(licium)* towers taking barges from Ostia to Rome. Though found in different sizes across the Empire, *codicaria* would nevertheless have carried a significantly smaller load than the Blackfriars 1 style ships, and would have been suitable for riverine use only.

These common types of vessel would have been joined by other ubiquitous designs in the Rivers Medway and Thames, for example smaller monoreme *myoparo* and *scapha* (types of cutter and skiff, Mason, 2003, 142). However these would not have fulfilled a function in the carrying of extractive materials and so are not considered here.

As with the transport of Wealden *Classis Britannica* tiles and Folkestone-region quern stones, the ragstone industry of the occupation-period upper Medway Valley would have made use of the regional river networks to take its part-worked merchandise to their respective places of use. In the case of London this was a journey that would have started upriver of Maidstone (the specific means by which the quarried stone was carried to the tidal reach is discussed below in Section 4.5.3), with the vessels then journeying downriver from the tidal reach, through the Medway Estuary to the Thames Estuary (see discussion in Section 4.5.3 below regarding a possible short cut to bypass the Isle of Grain), and then upriver along the River Thames to reach wharfing in London. Using the Blackfriars 1 vessel as the means of transporting the stone, Marsden (1994, 83) calculated that to construct the land walls of Roman London 1,750 such voyages would have been needed to carry the required 45,000 tonnes of ragstone for the facing alone. From the quarries above Maidstone this would have been a journey of some 127km each way given the convoluted nature of the route if no short cut was used (again, considered in detail in Section 4.5.3, see Figure 46)).

Looking at similar regional journeys carrying the extracted ragstone, Pearson (2002a, 82) details that 870 boatloads of the material had to be transported from the Medway quarries to the Saxon Shore fort of Bradwell to facilitate its construction, with 530 for Reculver (the latter a very good example showing demand for the ragstone given it was transported 50km further than any other material used). Both journeys would once more have begun in the upper Medway Valley, with the vessels making their way to the Thames Estuary. Those heading to Bradwell would then have crossed to the Essex coastline, following it northwards before entering the estuary of the River Blackwater where the fort was situated.

Meanwhile those travelling to Reculver would have turned right upon reaching the Thames Estuary, then tracking the coast before reaching the fort. Any vessels travelling on to Richborough or Dover would have continued east before slipping through the Wantsum Channel to reach their goals. Similarly any stone travelling to Canterbury would have entered the Wantsum before travelling up the River Great Stour to the *civitas* capital. Meanwhile the ragstone extracted for use in Colchester would again have travelled down the Medway, across the Thames Estuary, along the Essex coastline and finally up the River Colne.

By way of analogy, there are other examples of the long distance maritime transport of extracted materials in occupied Britain. For example Parker (2009, 61) points to the largest (3.65ha) Saxon Shore fort at Pevensey where he believes 1,600 long-distance boatloads would have been required to transport the building materials (in this case largely Greensand and flints, Pearson, 1999, 103). In similar examples, Hayward (2009, 112) points to distances of over 200km in the supply of freestone for memorials, while Allen and Fulford (2004, 30) add the distribution of mosaic tesserae as another example (they saying that the same small range of materials were used 'almost everywhere' from Kent to Devon and south east Wales). They also detail examples of the even longer distance maritime transport of building materials, for example the lava blocks from the Eifel Mountains in north eastern France used in the walls of the Saxon Shore fort at Walton. Jones and Mattingly (1990, 220) go into even greater detail regarding the building materials used in Roman London, citing exceptionally long-range examples such as black and white aquitanic marble from northern Spain, Carrera marble from Italy and Red Porphyry from Egypt. In this regard, Russell (2013a, 11) explains that such very long-range distances are disproportionately more common for these finer quality stones used for decorative purposes than for more familiar building stone.

Even in the latter context however, it is noticeable how much further ragstone was transported when compared to other common building materials during the occupation (see Reculver example above), especially given the particularly difficult journey when taking into account the complexities of transiting from the Medway Estuary to the Thames Estuary (see Section 4.5.3). I have therefore created Table 4.1 showing where Kentish ragstone from the upper Medway Valley was used in construction projects during the occupation, with all materials being transported by river and sea. It specifically shows the distance as the crow flies that the stone was transported, the chronology of its use at the location (note that where this shows a range into the 4th and 5th centuries AD this denotes continuous, local or re-use) and any relevant context. The sites are listed in this case alphabetically.

4.5.2 Was Riverine Hydraulic Infrastructure Used in the Medway Valley?

The use of waterways in a controlled fashion to support heavy industry and secure transportation across the Roman Empire was very common. As Ellis Jones (2012, 86) says: *"...Roman military engineering was more than capable of improving and maintaining rivers...establishing an effective series of inland waterways."*

A good example can be found around the intense Roman industrial landscape at Rio Tinto in south western Spain (discussed above in 4.4) where the Rio Guadiana and Rio Odiel were heavily utilised in a manipulated fashion to support regional silver, copper and lead mining activities (Jones, 1980, 148). Staying in Spain, similar high levels of hydraulic engineering skill have also recently been revealed by the use of LIDAR at the north western Spanish gold mining site at Las Médulas in the Eria Valley (Fernando-Lozano, Gutierrez-Alonso and Fernandez-Moran, 2015, 359). Here, an extensive network of man-made hydraulic channels and reservoirs (some of the latter with a capacity of 5,000m^3 of water) were created by diverting regional rivers. Campbell (2011, 117) provides a further example of riverine manipulation, detailing an inscription from *Axima/ Forum Claudii Ceutronum* dated to AD 163 describing how Marcus Aurelius and Lucius Verus: *"...restored the roads through the territory of the Ceutrones, which had been torn up by the force of torrents, by shutting out the rivers and leading them back to their natural bed and placing embankments in numerous places..."*

Yet another example of such engineering skills is provided by the 37km long Corbulo Canal (*Fossa Corbulonis*) in The Netherlands, a major hydraulic undertaking linking the *municipum Cananefatium* (later *Forum Hadriani*, modern Vooburg) with the Rhine at Leiden to the north and the Meuse/ Mass estuary to the south.

These engineering skills have their origins in Rome's initial contact with the Hellenistic east. Here, its engineers would have refined their riverine and riparian

Table 4.1: Distribution table for upper Medway Valley ragstone

Location	Distance	Chronology	Context
Bradwell-on-Sea, Essex.	59km	3rd and 4th centuries AD	Building material in the Saxon Shore fort.
Boughton Monchelsea, Kent	Within zone of extraction	1st through 4th centuries AD	Villa and remote bath house.
Canterbury, Kent	42km	1st through 4th centuries AD	Town walls, public buildings, built environment.
Colchester, Essex	75km	1st through 3rd centuries AD	Claudian temple, circus.
Dover, Kent	60km	1st through 4th centuries AD	*Pharos*, built environment.
East Farleigh, Kent	Within zone of extraction	1st through 4th centuries AD	Villa and temple.
East Malling, Kent	Within zone of extraction	1st through 4th centuries AD	Villa.
Eccles, Kent	7.5km	1st through 4th centuries AD	Villa.
Faversham, Kent	30km	1st through 4th centuries AD	Hogs Brook aisled building, Stone-by-Faversham temple/ mausoleum.
Keston, Kent	34km	1st through 3rd centuries AD	Cemetery.
London	50km	1st through 4th centuries AD	Forum, basilica, Governor's Palace, amphitheatre, temples, wider built environment, 3.2km land wall circuit.
Maidstone, Kent	Within zone of extraction	1st through 4th centuries AD	Villas.
Plaxtol, Kent	13km	1st through 4th centuries AD	Used as a building material in the three Roman villas here, as a platform in a funerary context and in a trackway.
Reculver, Kent	52km	3rd and 4th centuries AD	Saxon Shore fort.
Richborough, Kent	60km	1st through 5th centuries AD	Early built environment, monumental arch, later Saxon Shore fort.
Rochester, Kent	15km	1st through 4th centuries AD	Public buildings, built environment, town walls.
Snodland, Kent	10km	1st through 4th centuries AD	Villa.
Southfleet, Kent	21km	1st through 3rd centuries AD	Walled cemetery.
Springhead, Kent	22km	1st through 4th centuries AD	Temples, built environment.
Swanscombe, Kent	24km	1st and 2nd centuries AD.	Non-villa settlement.
Teston, Kent	Within zone of extraction	1st through 4th centuries AD	Villa.
Teynham, Kent	24km	1st through 3rd centuries AD	Used as the principal construction material in the Octagonal bath house at Bax Farm.
Thurnham, Kent	8.5km	1st through 4th centuries AD	Villa.

After Merrifield (1965), Marsden (1994), Pearson (2002a and 2002b) and Blanning (2015).

management skills to enable the control of waterways, and were soon transferring techniques learned on the Nile, Tigris and Euphrates to the Tiber in Italy and later into Gaul and Spain. Thus, by the time of the occupation, their engineering skills in this area were highly developed and were to be henceforth found manifest in Britain, with Ellis Jones (2012, 86) saying: *"...The Romano-British period saw the development of a coastal and riverine transport infrastructure that was not equaled until the advent of the Canal Age..."*

Such manipulation in a British context was controversially suggested by north eastern archaeologist Raymond Selkirk in his *The Piercebridge Formula*, published in 1983. His hypothesis, never proved, suggested that the Roman military deployed in the north of England were supplied principally by waterways such as the Tyne and Tees which were being physically controlled by the authorities using locks and weirs, and not by road. This control, he argued, allowed the rivers to be navigable much further up-stream than was naturally the case.

The principal academic rebuttal of Selkirk's theory was that of Anderson (1992, 212) who reasoned that, because of the nature of the rivers in the North East (which feature troublesome shoals, incised currents and river cliffs, and are prone to strong tides and high winds), navigation by medium sized boats, barges and ships would have been problematic upriver of Wallsend to the tidal reach, and impossible thereafter. Bidwell (1995, 395), in his review of Anderson's work, was even blunter in his criticism. He said that Selkirk's work was fundamentally flawed as, in his view, the easternmost bridge at Piercebridge was definitely that and not riverine hydraulic infrastructure as Selkirk had argued. He then went on to say that other locations identified by Selkirk as being occupation-period hydraulic river infrastructure were either later in date or natural.

That the Roman occupiers in Britain had the engineering capability to build such riverine and riparian infrastructure is not in doubt however. A good example is provided by Car Dyke in East Anglia where, at regular intervals along its 122km length, gravel causeways submerged to a depth of 1m and featuring timber pile foundations have been laid to facilitate crossing points linked to key local trackways (Catling, 2014, 31).

With such examples in mind, when finalizing the synopsis for my UCL Archaeology MA dissertation, I decided to use a template based on Selkirk's contested ideas for the north eastern rivers to see if the River Medway, with its broad flood plain and gently sloping banks, would be a better candidate for the use of occupation-period riverine infrastructure to allow its use to be managed in an industrial context.

As a starting point for what I called the 'Medway Formula' I looked at anecdotal evidence of activity upriver of the tidal reach where (as I demonstrate in Sections 4.1.4 and 4.3) there existed a thriving industrial landscape of ragstone quarries together with the villa estates where the controlling elites lived, all making great use of the river. In that regard, the key fact to consider is that in the period of the historical record the Medway was un-navigable above the tidal reach at Allington until the advent of the Medway Navigation Company in the 18th Century (earlier attempts in the 17th century by the Commissioners of Sewers being less successful). Referencing the River Medway back to the occupation, Kaye (2015b, 232) adds that when the Romans arrived in AD 43 the river was equally unnavigable above the tidal reach. Medway expert Ray Chitty (pers. comm. 07 October 2012) explains why:

> *"If you removed all man-made weirs, sluices and locks then the river would be little more than a stream. It is the locks and weirs that hold back the water and maintain depth. Specifically, I do not think the Medway would be navigable past Allington. And remember the sea level was lower in Roman times, so the tidal reach would have been even further downriver than today (at Snodland as argued above)."*

Thus, without river infrastructure, access to the ragstone quarries and villas where the elites lived would have been limited to those on the tidal reach at Allington. A useful analogy is provided by the Tiber, where before hydraulic river infrastructure was installed Pliny the Elder (III.53-4, 1940) describes the upper reaches as only being accessible to rafts and log boats. That certainly was not an option in the upper Medway where I contend Blackfriars 1 vessels and various *codicaria* were plying their trade supporting the ragstone quarrying industry.

Until recently such analogy, compelling though it is, was the only indicator that riverine hydraulic infrastructure was used on the upper Medway during the occupation. This is now beginning to change, with hard archaeological data emerging to support such a hypothesis based on research by the author. In the first instance, references have come to light in the historical record of riverine hydraulic infrastructure in the upper Medway being removed in the 18th century. Specifically, *'Mr Coles Observation of Nuisances on the Medway'* (found by the author within a late 18th century reproduction of the original 1627 to 1630 Proceedings of the Commissioners of Sewers, in the Medway Navigation Company's 1800 papers housed in the Medway Archives and Local Studies Centre, dated 15th June 1630) goes into great detail about stone weirs and shelves being removed from the river at East Farleigh, Barming and Teston (1630, 134). Though the Roman provenance of these is clearly unprovable today, it is no coincidence that all the locations are the site of historical river crossings on the Medway (each having a bridge today), implying that roads or tracks of ancient origin led to these points. Geographically, each site also has an association with activity during the Roman occupation, for example the villa at East Farleigh, the villa and ford at Barming and the villa at Teston.

Further, as is again detailed above in Section 5.1.4, the capping stones on the church wall at St Mary's in East Farleigh are large, clearly marine weathered and resemble similar material used in Roman river infrastructure elsewhere in Britain (for example the Roman ford at Barnard Castle on the River Tees in County Durham). A reasonable case can be made that these stones are actually re-used survivals from the removed weir/ stone shelf at East Farleigh.

Using Coles' observations regarding stone weirs and shelves being removed from the river as a guide, I next carried out a close study of the river in the 1797 Ordnance Survey maps of Maidstone and Addington to see if infrastructure was still visible in the River Medway, having survived the attention of the Commissioners of Sewers in the 17th century. Starting with Teston, visible adjacent the northern bank of the Medway to the immediate west of the Medieval bridge can be seen a long island, analogously very reminiscent of Jones' River Severn weir bypass barge-gutters (discussed in the context of Maidstone in Section 4.1.4). Meanwhile, and possibly linked to Coles' observation at Teston, local historian Severn (1975, 7) additionally talks of:

> *"...the remains of a Roman way which appears to have crossed the river at or near Teston and continued over Coxheath, (and which) was supposed to have been discovered by Dr Robert Plot (who lived in the 1600s)...there is in fact a crossing over the river bed near the bridge."*

Moving downriver, as detailed in Section 4.1.4, one next meets the island to the immediate west of East Farleigh Bridge, the weir between East Farleigh and Tovil and the two enigmatic islands parallel in the river at Tovil. Finally there is the now removed island in Maidstone detailed above.

Regarding the riverine hydraulic infrastructure itself, in the case of locks both Campbell (2012, 225) in his broad review of occupation-period riverine manipulation and Selkirk (1983, 94) in his controversial work argue that the most likely lock type used in Britain during the occupation would have been the simple flash lock, effectively a lifting gate in a weir (though a bypass channel could have been used but featuring only one gate). Such flash locks are ancient in design and well documented on the upper reaches of the Tiber (Selkirk, 1983, 94). Pliny (Epistulae 10/41-42 and 61-62) details the use of such technology in the scheme he proposed to the Emperor Trajan to build a canal to connect Lake Sophon in Bithynia to the Sea of Marmara. More recently and in Britain, Straker (1931, 189) records similar types of lock (called locally in this case 'shuts') being used to make the River Rother in East Sussex navigable in the medieval period. It is possible that more complex pound locks were also used during the occupation, utilising two lifting gates to facilitate the lock's use and also featuring a bypass channel (effectively a modern lock), and indeed Moore (1950, 97) has argued that it is this technology which Pliny describes above.

While considering the use of riverine hydraulic infrastructure in the Medway during the occupation, and indeed the types of technology likely to have been employed, it is also important to consider the legal framework within which they were built and controlled. Such waterways were obviously vital not only for transport purposes, but also to provide a water source for agricultural purposes (both in the riparian zone and indeed the wider landscape). This will have been especially important given the agrarian nature of the Roman economy. Campbell (2011, 87) has made a detailed study of the available legalistic historical documentation on waterway ownership and says:

> *"Effectively, rivers belonged to the Roman people, were part of public property, and therefore were accessible to all. However...ownership of the banks belonged to those whose landholdings was contiguous, and they also possessed any trees on the banks. In fact, the riverbank was public property only from the point where the banks sloped down toward the water."*

The river itself was therefore owned by the state representing the Roman people, being dubbed the *'flumen publica'*, though it is not clear as to the level of 'state' engaged here, possibilities ranging from the provincial Procurator at the top to local magistrates at the bottom. Private individuals controlled the banks. A useful analogy here explaining how (within this framework of the state owning the river itself) manipulation of its course could be facilitated is provided by the *municpaeum lex Irnitana* in Spain which awarded grants to local magistrates to create or alter the course of waterways (Campbell, 2011, 89). This was a case of local Government responding to the needs of the local community, though a reasonable argument can be made that for extensive riverine hydraulic infrastructure to support state-run industries such as the ragstone quarries of the upper Medway Valley, the Imperial Government would have been directly involved. A further analogy in this regard would be aqueducts, which were controlled directly by the Emperor, with Campbell (2011, 91) saying: *"The Emperors were masters of the aqueducts, through control of their finance and administration and the power to take initiatives. This included money for aqueduct repairs."*

An interesting point here concerns the legal distinction between a river (*'flumen'*) and a stream (*'rivus'*) in Roman law. Campbell (2011, 88), again citing extensive historical research, explains that the application of either definition for each individual waterway was actually based on the local opinion of those using, working near or living along the river. This raises the interesting prospect of the definition of the upper Medway being changed during the occupation from stream to river once the riverine hydraulic infrastructure was installed to facilitate access to the industry and settlement based there.

Finally, when considering the legal framework for river manipulation during the occupation, Ellis Jones (2012, 6) makes an interesting comparison between the heavily legalistic experiences of those endeavouring to open up the Medway from the 17th century onwards and their Roman forebears. The latter were simply faced with a virgin river lacking any previous infrastructure which was ready for their use and improvement, while the various iterations of the Commissioners of Sewers and the Medway Navigation Company had to contend with the physical manifestations of all preceding activity in the river (especially from the Roman occupation).

4.5.3 What Would a Typical Journey Look Like?

Having considered the River Medway itself, the distribution of ragstone from the upper Medway Valley and the use of riverine hydraulic infrastructure to facilitate the transportation of the extracted material, we can now consider what a typical journey for a boatload of ragstone from quarry to its place of use would have looked like (in the case of this example London).

Prior to looking at this journey in detail two natural factors are first considered here which played a major role in the success of the journey, namely the tides and seasonality. In the first instance, crucial to the journey's completion was a detailed knowledge of the local tides. It is the gravitational attraction of the moon and to a lesser extent sun (the closeness of the former countering the far great mass of the latter) on the rotating earth that creates tides and thus influences where the tidal reach is. In Britain today there are usually two high tides (every 12 hours and 25 minutes) and two low tides per day, this being called a semi-diurnal regime. In addition to this twice daily rise and fall, the specific positioning of the moon then also

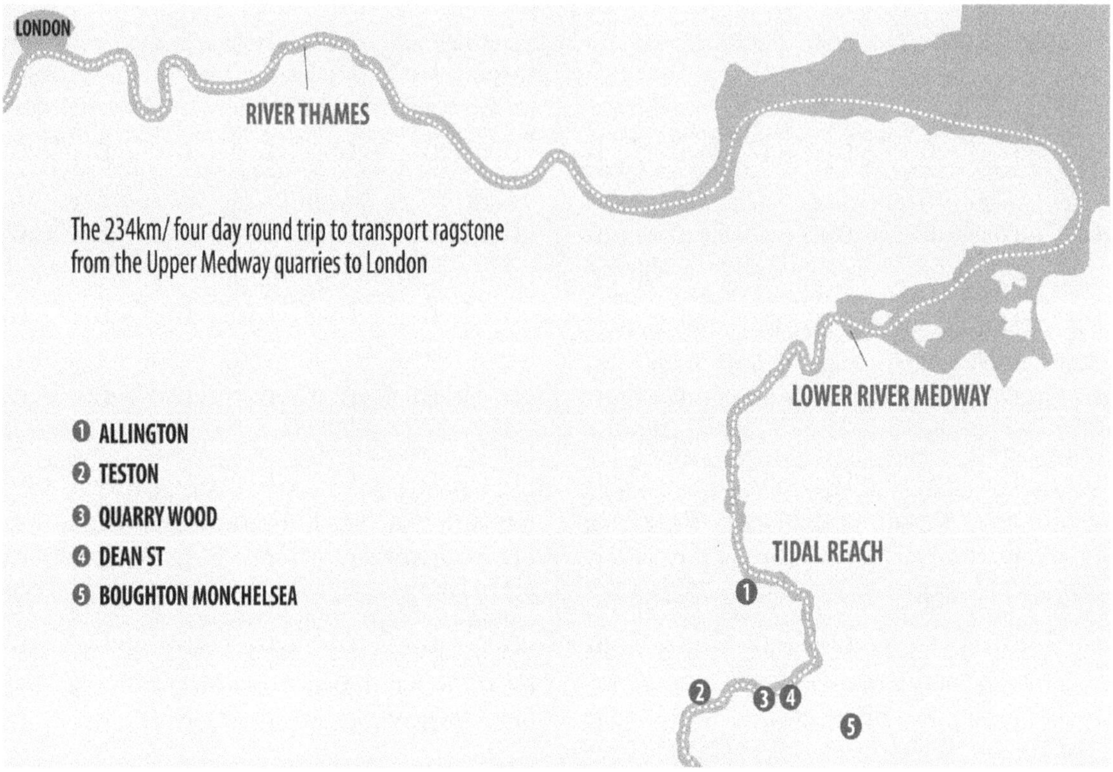

Figure 46: Stylised view of the 254km round trip of a boatload of Kentish ragstone from the upper Medway Valley to London, from the River Medway, round the Hoo peninsula and down the Thames Estuary. Paul Baker.

provides additional influence such that the highest high tides and lowest low tides are spring tides (with full and new moons), and the lowest high tides and highest low tides are neap tides, this being a fortnightly cycle. Finally, a seasonal cycle also has an impact such that the highest of the spring tides peak during the spring and autumn equinoxes. Within this complex tidal regime the difference between high tide and low tide is known as the tidal range, this additionally differing around the coasts of Britain based on regional littoral and riparian topography, and also water depth (bathymetry).

Our next consideration is that of seasonality and the use of the River Medway during the occupation. The key impact here would have been caused by the weather itself, not least precipitation levels. Modern scientific research indicates that during the Roman period weather conditions were not far removed from that we experience today, excepting it was perhaps slightly wetter earlier in the occupation (Grainge, 2005, 37). Changes of season would have specifically affected the use of the Medway to transport heavy goods such as ragstone in that out-of-season sailing conditions in the Thames Estuary and around the coast would have impacted the flow of traffic on the river. Ellis Jones (2012, 24) points out that gales of Force Seven and higher are eight times more frequent in the waters around Britain in the winter months, and even in the occupation period itself the dramatic impact of worsened sea states on the inland river systems of Britain was well known (Tacitus, 1970, 61). The likelihood is that out-of-season coastal traffic would have continued, and therefore the need to use the river, but at a lesser rate determined by the weather. If conditions were very poor, vessels would simply not have begun their journeys. Of course, if conditions were severe enough, for example during an acute cold snap, the river itself may have actually frozen, thus totally preventing its use. Analogously, Ellis Jones (2012, 29) records this occurring on the River Severn in January and February AD 1695, and in the winter during AD 1716-17. Meanwhile, at the other extreme, severe drought would also have hindered riverine travel during the summer months (even with 'Medway Formula' infrastructure in place). Another factor to consider here is that of flooding. This is specifically relevant above the tidal reach, where the river level itself is only affected by rainfall and ground drainage. In the modern era the effects of flooding during unusually heavy raining or snowing events has been minimized by the construction of the Leigh Barrier above Tonbridge which provides additional protection against flooding. During the occupation though, lacking such technological sophistication, the river would have been more prone to flooding, although hydraulic riverine infrastructure would have gone some way to mitigating against this.

Taking all of the above into consideration, one should envisage the River Medway as being a thriving, busy waterway during the Roman occupation excepting during extreme weather events and perhaps in the depths of winter. Using the anecdotal imagery provided by Ausonius (Mosella, 3-16), in normal circumstances one can imagine the *myoparo* and *scapha* as they skimmed up and down the river while Roman 'hufflers' laboriously pulled their *codicaria,* with Blackfriars 1-style vessels heeled over into the wind carrying their loads of ragstone to London.

With regard to these vessels and larger ships, and their association with the ragstone quarrying industry, we are faced with three possibilities regarding how the stone quarried upriver of Allington actually arrived at the tidal reach for its journey onwards (I have ruled out any form of portage of the vessels given the weight of their loads).

The first option is that the quarried stone was transported by cart from the upper Medway quarries over Malling and Barming Heath and on to the tidal reach of the river, there to be loaded onto Blackfriars Type 1 vessels. This would have cut out the need to traverse the non-tidal Maidstone bend upriver of Allington. Certainly both Allington and Aylesford have historical associations with the operation of busy quaysides, and the Roman villa at East Malling is almost equidistant between Teston/ Barming and Aylesford, so a reasonable case can be made that this villa's location was associated with facilitating industrial transport if a roadway can be located linking these two points. Aside from such anecdotal evidence however, common sense seems to mitigate against this option. As many of the commentators quoted throughout this research have said, the preference would have been to put the ragstone onto maritime transport as soon as possible given it was the most ergonomical solution by far. Specifically, Pearson (2006, 91) says: *"Economic logic...suggests that the movement of heavy goods over long distances by land was a last resort, occurring only when navigable water was not accessible, or its use impractical."*

Additionally, a prominent haulage road would have been required and none is evident. Hasted (1797, 2) does describe a Roman road running from Maidstone through Barming and on to Ightam, though hard data supporting this is conspicuously lacking and it is also on the wrong alignment to traverse the Medway bend. Additionally, snow marks indicating trackways across Malling Heath were noted during a 1970s MAAG survey, but again no substantial archaeology was found.

The second possibility is that those operating the quarries transported the ragstone loads by *codicaria* from the upriver quarries to a break-of-bulk point riverine port above the tidal reach, it then being transferred onto Blackfriars 1-type ships. As detailed above in Section 4.2, such a transfer in this scenario was essential given the unsuitability of the towed river barges for operation in the Medway and Thames estuaries. This break-of-bulk point solution is supported by Ellis Jones (2012, 101), with Rochester, Allington on the tidal reach (or indeed Snodland during the occupation as argued by Kaye 2015b, 232) and the area of modern Maidstone being potential candidate locations. If Rochester was the site of such a break-of-bulk point riverine port, then this would also have precluded the need for a drawbridge at the Watling Street crossing of the Medway if larger vessels were not being used upriver of the small town. Wilkinson (2006, 14) points to regional Medieval parallels of such transhipment from smaller to larger vessels, for example at Hollow Shore on the north Kentish Coast where Faversham Creek meets The Swale. Here barges were employed to carry goods up Faversham Creek where they were then transhipped onto sailing vessels for onward transport.

Having travelled on the Medway between Allington and Teston however, I am inclined to endorse a third option, the one ship solution. This features a Blackfriars 1 type vessel completing the entire ragstone-carrying journey, with *codicaria* instead being used to support local agriculture and to transport ragstone for local use (Wilkinson, 2006, 8, suggests that such towed barges were used for transporting agricultural produce from large villa estates during the occupation in Faversham Creek). To my mind it makes no sense ergonomically to tranship loads at potential riverine port locations such Rochester, Allington or Maidstone given the short distances involved (the Medway is not a major continental river such as the Rhine for example). Further, the loads of ragstone would have been significantly larger and bulkier than those being transhipped at Faversham in the medieval period (see Wilkinson quote above). Additionally, the width of the Medway downriver of Allington, and particularly downriver of Snodland, would make towing *codicaria* from a towpath problematical, even using the advanced techniques recorded for the Rhine and Moselle. A reasonable argument can of course be made that as the Medway is as shallow as 2m or less in places upriver of Maidstone (I have experienced this myself when travelling on the river), then the use of some types of sailing vessel would have been problematical, but the shallow carvel-built draft of the Blackfriars 1 design would be ideal in such conditions. Indeed, Marsden (1994, 80) has postulated that: *"The last voyage of the Blackfriars ship was...down the winding River Medway from the Maidstone area, past the Roman town of Rochester, and into the Thames Estuary."*

Sticking with my preference for the one-boat journey, and using archaeological data and my knowledge of the Rivers Medway and Thames, one can perhaps build a picture of a typical journey for a Blackfriars 1 type ship from the upriver Medway quarries to London and back. This is predicated on three key points, these being:

- As I argue above, using a break of bulk point would have been uneconomical so one type of vessel would have been used for the whole journey.
- The vessel would have to be capable of getting under sail with a 50 tonne load of ragstone, but with a shallow enough draft to operate upriver of Allington on the Medway.
- Such a vessel would have had to be able to navigate the 'Medway Formula' river infrastructure which I argue made the Medway upriver of Allington navigable in the first place.

Taking this into account, I believe that a trip would look something like the below. It would have taken two days to make the 127km journey if it were unbroken excepting an overnight stop, though note there is no reason the vessels would not stop off along the route to drop off their mercantile wares and pick up others:

- Load (of finished stone, rough outs or rubble) taken aboard a Blackfriars 1 type vessel at Teston, East Farleigh, Tovil or Maidstone. Marsden (1994, 80) and Merrifield (1965, 49) are clear that the wharfing used to load the ragstone would have been as close to the quarries as possible (see Section 4.1.4 regarding the possible such wharfing for the Dean Street quarry found in association with the 'Medway Stones').
- Ship uses sails, towers or rowers, and the flow of the river, to get to the tidal reach at Allington. With regard to sailing, the predominantly southwesterly winds would have allowed its square sail as detailed by Marsden in Appendix C to be utilised, noting that the use of such sails above the tidal reach would have required the vegetation on both sides of the river to be cut back considerably. Ausonius (Mosella, 7) is insightful here, speaking of the banks of the Moselle being cleared of vegetation and covered in hard, compacted sand (I do note that the Moselle is clearly a broader river than the Medway, but believe the analogy still stands). This would also have facilitated towing, with Wilkinson (2006, 14) anecdotally describing the medieval practice of towing vessels with loads of up to 80 tonnes the three miles from Hollow Shore on the Swale to the dockside at Faversham, with rowers being used if saving money was a necessity. Ellis Jones (2012, 91) similarly points out the use from the early medieval period of 'bow-hauliers' on the Severn when conditions prevented the use of sails. In a Medway context, such towers would also have assisted navigating the occupation period 'Medway Formula' hydraulic river infrastructure. Wilkinson (2006, 14) adds with regard to using the flow of the river that if this were too swift, perhaps after heavy rain, then a capsize anchor would have been used to control the speed. To cover the 8km from Teston the speed would have been around 2 knots and the journey would have taken up to 3 hours, including the use of any riverine hydraulic infrastructure.
- Having reached the tidal reach of the Medway at Allington (or Snodland during the occupation, Kaye 2015b, 232), after which the width of the river and the tides would have precluded the use of towers, the vessel would have waited for a falling tide to navigate the tidal section of the river up to Sheerness, the total journey of 45km taking upwards of 6 hours assuming a reasonable wind and a speed of around 4 knots. Given that the distance between the tidal reach and the Thames Estuary is greater than it was possible to navigate with one tide, even a high spring tide, the use of sails would have been essential. Anecdotally here, having extensively sailed in the Medway myself, it is clear to me that to navigate the lower reaches of the river successfully the boat crews would have needed an exceptional knowledge of the local tides and winds, especially given that their sail technology would have been deficient when compared to modern examples (again see Appendix C below). The vessel would then have entered Morris' southern North Sea and Eastern Channel connectivity system (2010, 10), or Evans' East coast trade route (2013, 433).
- By this point, even in summer, the vessel would have been struggling with the light and it is inconceivable any journey would have been attempted from this point without reasonable visibility. Therefore it seems likely that the vessel would have used an overnight anchorage before continuing the journey the next day. Sheerness seems the most likely candidate, though an interesting option here is presented on pre-modern maps which show a creek called The Dray (interestingly, Old English for 'to pull') isolating the Isle of Grain from the Hoo Peninsula. This is now silted up and, given the balance between rising sea levels and silting, it is unknowable if this creek existed in Roman times or not. If it did though, it would have provided ideal shelter for this overnight section of the journey and would also have provided a short cut to avoid travelling to Sheerness before entering the Thames Estuary. Kaye (2015a, 29) goes further here, showing that the Isle of Grain could have been even further detached from the mainland during the occupation as a full (and diminished in size) island, presenting even more possibilities for an overnight stay.
- The following morning the vessel would again wait for the early tide before sailing up the River Thames to London, a journey of 74km which at 5 knots would take around 8 hours.
- Travelling back, perhaps with exotic goods and pottery (for example south Essex and north west Kent sandy grey wares, Houliston, 1999, 163) for the elites and artisans living along the north Kent coast and Medway Valley, the vessel would have used its sails, the current and the tide to reach the River Medway. Specific insight for the types of exotic goods carried comes from the work of Andrews *et al* at Northfleet (2011, 223). Here, archaeological evidence has been found of amphora carrying fish products from Portugal and olive oil from Spain and North Africa at a villa estate with a quayside on the Ebbsfleet. One should note here also the remains of Spanish amphora found at the large villa site at Teston (see Section 4.1.4).
- Following once more an overnight stop on the Isle of Grain or Sheerness, sails and the tide would then have then been used to travel upriver to Allington, with a number of vessels perhaps being tied together.
- Finally, above the tidal reach again, towers or rowers would again have been employed to facilitate travel along the short distance to the ragstone quarries while navigating the 'Medway Formula' hydraulic river infrastructure.

5

Discussion – Interpreting the Evidence

In this work a number of issues and concepts have been newly revealed for the first time. In the first instance, at a general level, the hitherto unacknowledged depth to the economic exploitation of natural resources by the two *metalla* enterprises considered in Kent and the South East during the occupation, and their role as part of the Imperial economy. For the iron industry in the Weald through to the mid-3rd century AD, the research has then specifically shown for the first time the division in terms of geography and scale of the central and the eastern/ coastal iron manufacturing zones. Meanwhile, in the upper Medway valley the research has for the first time fully illuminated the large scale ragstone quarrying industry there which provided much of the building material for the region during the same time period. This has included locating the five specific quarries utilised by this *metalla* for the first time.

The research threads that have emerged from this extensive research are far too numerous to consider in one volume. I have therefore decided to focus this concluding discussion on two broad themes which I think provide the most original insight and value, updating and in some cases transforming many existing views and received wisdoms. These themes are:

- The role of the state in Roman industry, in the case of this research the *metalla* mining and quarrying industries actually exploiting the natural resources in the region. Many commentators (all detailed below in Sections 5.1.3 through 5.1.4) have speculated, both from a well and an ill-informed perspective, that the state was a prominent feature of these industries, often in the form of the *Classis Britannica* regional navy. Some have gone even further and set these industries within the context of a regional Imperial Estate. Such assertions will be tested here based on data and interpretations made newly available in this work, this contribution being particularly valuable given the lack of detailed analysis of such state-controlled industries and estates (Carroll, 2016, 33).
- The events of the mid-3rd century AD when, as I have demonstrated above, great change took place in the political, economic and social outlook of the Roman south and east of Britain, as elsewhere in the Empire. This discussion is framed in the context of change and continuity regarding the exploitation of natural resources by the extractive industries in the region during the occupation (together with associated settlement and other supporting industrial activity). In that regard the section also incorporates, to provide context, consideration of both the LIA/ occupation transition and the end of the Roman experience in the South East.

In terms of methodology, both themes are broken down into sub-sections to allow for an appropriate level of forensic interrogation of the evidence by the reader. Under each theme these sub-sections include two specifically dealing with the two regions of interest within the research, with the same process being used for each in that the existing data and interpretations are considered before the new data is reviewed and analysed. This new insight is then used to suggest any transformation of existing understandings. At the end of each of the two themed sections I conclude with a synthesis to provide a final commentary.

5.1 The Role of the State

In the debate about the nature of the Roman economy outlined in Section 2.2.1 I set out the arguments in favour of the view that the Roman economic system was in many ways similar to that of the western pre-modern economy (this view supported by the 'modernists') and the arguments that it was less so (supported by the 'primitivists'). One area however where there were clearly parallels was with intervention by the state, often in the form of the military, to carry out tasks to support political, economic and social institutions. Examples include administration, engineering, construction, and of particular relevance to this research the management (and in some cases full operation) of major industrial enterprises such as the larger mining and quarrying *metalla*. The latter are of significant importance to this research given its focus on natural resource exploitation in occupied Kent and the South East. To that end this themed section of the discussion is devoted to an examination of the role of the state in these industries in the Weald and upper Medway Valley.

In that regard, as set out in Section 2.2.1 above, it is worth reminding ourselves of the various means by which the Procurator (through his *procurator metallorum*) could manage the large-scale extractive industries within his provincial remit. Mattingly (2006, 494) is clear in his view that the large-scale exploitation of natural resources was a key area of state control within the Imperial economy. Such state involvement indicates a degree of direct control by the Procurator (rather than indirect control through the provincial economy), and therefore the question to be asked is at what level was this control exercised.

At its most extreme form this direct control would have been in the form of an industrial Imperial Estate, the options for their management on behalf of the Procurator being:

- By the military, certainly in terms of opening up the opportunity but often on a longer-term basis.
- Through the use of procuratorial *vilici* (bailiffs).
- Through chief tenants (in the form of a head lease *conductores* for example), or under other forms of tenurial arrangement.
- To confuse matters, any combination of the above.

The other option to run large *metalla* industries which were not part of an Imperial Estate, but which still sat within the Imperial economy under the Procurator, was through the use of contractors or natives under production agreements and licenses with the office of the Procurator (though noting that even in this case such contracted-out *metalla* enterprises were often initiated by the military before being contracted on).

In order to provide definition in this review of the state's role in the *metalla* of Kent and the South East, given the wide range of options detailed above for such interaction, this sub-section will therefore firstly review the above research concerning large scale industry in the region to inform the ensuing analysis, then examine the types of role the state might have played, before discussing in detail the official presence in the industries of the Weald and the upper Medway Valley. The section will then compare the revealed nature of this state presence in regional industry with other manifestations of Roman officialdom in Kent and the South East to consider the level of military's role here, before concluding with a synthesis of the preceding discussion.

5.1.1 Large Scale Industry

As is evident from the regional studies and preceding background chapter, this research project has revealed a previously unrecognised depth to regional industrial activity during the Roman occupation. This has of course included the extractive industries exploiting natural resources, but also other industrial activity (related and otherwise). Here these industries are detailed to facilitate the following discussion in this section regarding state activity.

Starting at the upper end of the range, it is clear that through to the middle of the 3rd century AD Kent and the South East were one of the industrial heartlands of the wider province (Mattingly, 2006, 509), for example in the form of the iron manufacturing industry in the eastern/ coastal region of the Weald (larger for instance than the early iron industry in the Forest of Dean, Jackson, 2012, 169, and that in East Yorkshire, Halkon, 2011, 148) and its associated tile and brick manufacturing industry. The ragstone quarries of the upper Medway Valley provide an even more striking example, the combined quarry area for the five identified being 723,430m^2, a figure which has no parallel elsewhere in Roman Britain at any stage of the occupation (a case could be made in this regard for the quarries used to provide the material to build Hadrian's Wall, but as detailed in Section 4.3.3 the 11 known were much smaller, more localized and far more widely distributed, and so I would argue not part of one *metalla*).

Existing (Hodgkinson, 2008, 28, Jones and Mattingly, 1990, 217, and Pearson, 2002a, 82) and new (S. Elliott, 2013, 40, and 2014a, 251) data show these Kentish *metalla*, starting from a low base though clearly with a degree of foreknowledge of natural resource potential prior to the occupation, rapidly boomed. The expansion of the iron industry catered for demand created by the military expansion north and westwards; the ragstone industry for the urban expansion in the south and the east (and particularly London). Thus, by the late 1st century AD, massive industrial enterprises of the scale of Peacock's manufactories (1982, 8) were operating at full capacity in both regions. These two extractive industries were linked by the Rochester – Wealden roadway connecting the north Kent coast with the most significant iron working sites in the eastern/ coastal region of the Weald (this especially the case if the spur linking this route directly with the upper Medway Valley is confirmed, see Section 3.5 above and Figure 47)). As detailed in Section 3.1.4, a possible riverine link is now also being investigated given the potential quay at the Great Cansiron iron-manufacturing site in the central region of the Weald (Russel and Staveley, 2012, 1).

Both of these industries would have been the principal economic foci in their respective Wealden and north western Kent economic zones through to their demise on this scale in the mid-3rd century AD (Mattingly, 2006, 386, and see Section 5.2.2 and Section 5.2.4), with much of the wealth created being channelled directly through the *procurator metallorum* to the Procurator and then onwards to the Imperial *fiscus* (as part of the Imperial economy). In so doing they would have contributed significantly from an early date to ensuring that the province was *pretium victoria* – worth the effort of conquest.

5.1.2 What Role Could the State Have Played in Roman Industry?

To provide context for the three following regional discussions, here I briefly review the role played by the state in the economic institutions of the Empire, and the tasks performed by the military therein, starting with a broad focus but then concentrating specifically on industry.

The Roman state was not only the principal political force within the Empire, it was also intricately involved in economic activities geared to supporting its infrastructure, its military and its own continuance. Yet as a patrician institution it also had other responsibilities, with political, social and economic roles which were often related. A high profile example would be the supply of grain to the citizens of Rome (detailed above in Section 2.2.1). To give some idea of the scale of this commitment set within the wider economic system, it included in the reign of Claudius

Figure 47: Roman road near to the quarry alongside Dean Street in the upper Medway Valley. A possible spur from the Wealden Road which would then directly link the Wealden iron industry and upper Medway Valley *metalla*. Simon Elliott.

the construction of the new harbour 3km from Ostia to facilitate the growing capital city's demand for increasing quantities of grain (Erdkamp, 2013, 272). The imported grain itself arriving at this new port was often state-owned, originating from numerous agricultural Imperial Estates in many of the provinces of the Empire (increasingly from Egypt and North Africa, see Section 2.2.4), and was transported on shipping coordinated by the state (Erdkamp, 2013, 272, see Section 2.2.1). The significance of such agrarian Imperial Estates is ably illustrated by the extensive wine-producing Imperial Estate at Vagnari to the east of the Apennine mountains in ancient Apulia which was large enough to feature its own *vicus* (Carroll, 2016, 31). As detailed above in Section 4.5.2, the state also had responsibility for key public amenities, for example aqueducts (Campbell, 2011, 91), and also key elements of the transport network including the river and canal systems themselves (these being publicly owned, Campbell, 2011, 87). To this we can add of course the *cursus publicus* state courier service (Kolb, 2001, 95, and Burnham and Wacher, 1990, 5).

In an age before the advent of any modern civil service, nationalized industries and a modern free market to facilitate major capital expenditure programmes on behalf of the state, it was often the military who were deployed by the state to carry out tasks for which it was responsible (Mattingly, 2006, 523, and Elliott, 2016, 89). As set out in Section 2.6.3, a good example would be Imperial administration with for example a Governor's military *officium* being the principal tool used to both govern a given province but also being capable of turning its hand to any administrative task required of it (Goldsworthy, 2003, 144). D'Amato (2009, 14) also highlights the rather more practical task of urban firefighter for the military on behalf of the state. Clearly the military were also highly skilled in engineering, for example playing a leading role in the construction of the new harbour facilities at Ostia in the example detailed above (Erdkamp, 2013, 272). Military units (both land based and marine, the *Classis Britannica* in the case of the latter in Britain, D'Amato, 2009, 15), especially earlier in the Empire, were well serviced with specialist craftsman and engineers who would be the first choice resource to be deployed by the state for large-scale construction projects such as road building and built-engineering infrastructure (Goldsworthy, 2003, 146, see Section 2.6.3 for detail).

Looking specifically at industry, there are many examples of state control. This included responsibility, again often through the military, for running the major *fabricae* which provided equipment to all branches of the armed forces. Esmonde Cleary (2013, 93) illustrates the scale and sophistication of these manufactories, with for example that at Autun specialising in ballista, armour and shields while that at Reims specifically made swords. The state also controlled the official mints producing coinage for use across the Empire (Moorhead, 2012, 8), and in many cases managed the major brick and tile manufactories such as that detailed in Section 3.6 in the eastern/ coastal Weald (Mills, 2013, 461). The products of such brickyards were often officially stamped to mark their provenance and quality, in the case of the Weald with the CLBR symbol of the *Classis Britannica*. Topically for this research the state (again often in the form of the military) also managed the key *metalla* industries across the Empire, they being a key feature of the Imperial economy. This included the larger mining and quarrying enterprises, with Hirt (2010, 106), Cleere and Crossley (1995, 66), de la Bédoyère (1992, 100) and Jones and Mattingly (1990, 192) all arguing the majority of such operations exploiting natural resources were under state control through the military, ensuring the continuity of supply of the mined and quarried materials and the flow of wealth to the imperial fiscus. Specific examples in a British context are detailed above at 2.6.3.

Below I now specifically examine the case for such state control, through the military in the form of the *Classis Britannica*, of the extractive industries exploiting natural resources in the occupation-period Weald and upper Medway Valley.

5.1.3 The State Presence in the Weald

Many leading archaeologists and historians (detailed below) have long hypothesized that the state in the form of the *Classis Britannica* managed the iron manufacturing industry in the Weald during the Roman occupation. Modern research in this regard has focused specifically on the eastern/ coastal area where the largest sites were located and with its close association with maritime trade (see Section 3.4 for detail). Some commentators have taken this view of a state presence even further, making the case for the region being an official Imperial Estate (see Section 2.2.4, and below for detail). There is certainly a large amount of archaeological data pre-dating this research to support such views and I set that out here, followed by my newly presented data, to test this common hypothesis.

Adherents to the view that the regional navy managed the Wealden iron industry include Brodribb (1979, 141), Pearson (2002a, 50), Hodgkinson (2008, 33) and Harrington and Welch (2014, 109), while those going further in supporting the Imperial Estate theory include Marsden (1994, 83) and Cleere and Crossley (1995, 69, this based on Cleere's original work, 1977, 18), while Mattingly alludes to it (2006, 386). As those with an interest in the occupation-period Weald will attest, these individuals effectively represent the entire canon of those whose research has shaped our modern appreciation of Roman activity in the region (with the notable exception of Millett, see below).

The most frequently used existing data referenced by the above commentators linking the Wealden iron industry to the state and the regional navy is tile and brick stamped with a *Classis Britannica* mark (Brodribb *et al*, 1988, 275), this appearing in large numbers across the eastern/ coastal sites at Bardown, Beauport Park and Little Farningham and the associated port at Bodiam. In a specific example Brodribb (1979, 141) detailed that of the 41 complete *tegula* found in his excavations at Beauport Park in the 1960s and 1970s, all but one featured a *Classis Britannica* stamp. It is worth noting here that of the 3.35 tonnes of tile actually found on the site during these excavations (the vast majority of them only partial survivals), nearly all featured full or partial *Classis Britannica* marks (some 1,320 at the time, though this has since risen to 1,600). To further emphasise the scale of this occurrence, Brodribb (1979, 141) said that given that the overall size of the site examined was 114m^2, and the fact that the total number of *Classis Britannica* stamps found at the time was 1,320, this represented 11 such stamps per square metre. Such use of *Classis Britannica*-stamped tile to identify the nature of an occupation-period site has other regional parallels, with Philp (1981, 100) using the very dense concentration of over 1,000 such tiles at the original Roman fort at Dover as evidence that it was directly associated with the regional navy. One dissenting voice here with regard to using such tile to interpret the nature of occupation-period sites is that of Millett (2007, 178), who says of the Weald: *"The suggestion (based on tile evidence)...that a substantial part of the industry (here) was under the direct control of the fleet and that rights over iron were owned by the Roman state is entirely speculative."*

In his opinion the only exception here might be Beauport Park given the very large number of *Classis Britannica* tiles found here by Brodribb and others (Brodribb, 1979, 141, and Brodribb *et al*, 1988, 275).

There is further physical evidence however of a state presence in the eastern/ coastal region of the Weald, this being in the form of three pieces of additional epigraphy. In the first instance. and as detailed in 3.1.4, Brodribb *et al* (1988, 269) record their finding at Beauport Park of a wooden tile comb featuring the, to date, unique imprint of the letters CLBR, identifying this as a marked tool of the *Classis Britannica.* Next, at the same site, they (1988, 261) also detail the location of an inscription on stonework above the bath house entrance which references a *vilicus* (Brodribb *et al*, 1988, 261, see Sections 2.2.4 and 3.1.4). This is a most interesting term, originating as with Procurator in a domestic context (a *vilicus* being an agricultural estate bailiff), which by the time of the occupation was being used in a variety of official ways. One such was as a *vilicus officinae* tasked with managing an industrial enterprise, for example a mine or quarry, with Hirt (2010, 288) detailing just such an individual carrying out this role at the state-run Carrara marble yards in Rome, and in the same context Brodribb *et al* (1988, 241) referencing a similar individual running the state-managed iron-ore mines located in the Sana Valley in Bosnia. Both of these examples indicate an industrial Imperial Estate interpretation. While Millett (2007, 179) contests the state-association of the *vilicus* detailed in the Beauport Park example, to my mind it very clearly indicates the state representative who was responsible for the site during his time there, again a strong indicator of an industrial Imperial Estate interpretation (see below). The final piece of epigraphy is in the form of an iron die found in London which features a stamp declaring its provenance as *m(etalla) p(rovinciae) B(ritanniae)*, referring to its origins in a provincial iron manufacturing facility (Birley, 2005, 300). Sadly no direct link can be made with this and the Wealden iron industry, though it is easily the nearest major area of such official activity.

In terms of other material culture from the major sites in the eastern/ coastal zone which have been used to suggest a state/ *Classis Britannica*/ Imperial Estate presence, Henig (Brodribb *et al*, 1988, 260) argued that an *intaglio* featuring a representation of Victory found in the bath house excavations at Beauport Park also suggests the regional fleet being in attendance. Further, Hodgkinson (2012, 1) argues above in Section 3.1.4 that the Antoninus Pius medallion found by the IHRG in 2006 at Bardown is evidence of a significant state presence there, he saying that this token of prestige would have been the property of a high ranking official.

Those who have argued in favour of the state/ *Classis Britannica*/ Imperial Estate interpretation for the occupation-period eastern/ coastal Wealden iron industry also cite other evidence. For example Cleere and Crossley (1995, 62) detail the transport infrastructure of the occupation period eastern/ coastal Weald. They argue that the that the north–south alignments for the key regional Roman roads, particularly the Rochester – Wealden road to Beauport Park, speaks to their being built specifically to facilitate official communications for the *Classis Britannica* with the Medway Valley and the north Kent coast (though note the Lewes Road to London would have been equally important for the iron manufacturing activity in the central region). The location of port facilities at sites such as Bodiam to facilitate (it is presumed) the transport of manufactured iron out of the region also indicates the involvement of the *Classis Britannica* according to Hodgkinson (2008, 34, though one could argue it is likely the port also handled other exports such as wool and wood). The unusual settlement pattern of the occupied Weald, with little elite settlement in the centre and most of the villas on the periphery (with any Wealden elites also potentially residing in the Medway Valley, see discussion above in 3.5), also supports an industrial Imperial Estate interpretation according to Cleere and Crossley (1995, 58). The latter also highlight the rapidity of the expansion of iron manufacturing at the eastern/ coastal sites from the mid-1st century AD as evidence of a state presence (1995, 62).

Finally in terms of the existing data, Hodgkinson (2008, 34) has directly linked the disappearance of the regional navy after AD 249 with the similar decline of the vast majority of the iron working sites in the eastern/ coastal Weald in the same time period (particularly the larger Grade 3 and 4 sites).

Moving onto data and interpretations newly available with this research regarding the exploitation of natural resources in Kent and the South East during the Roman occupation, in the first instance we can consider numismatic evidence in the form of coin hoards. Specifically, this research considers in Section 3.4 for the first time (in an academic context in terms of its wider significance) the controversial 2008 High Weald coin hoard found at an occupation-period satellite iron working site near Bardown (the exact location being publicly undisclosed). This contained 2,891 radiates dating from AD 215 through to AD 268 (with all Emperors from Caracalla to the Gallic Emperor Postumus being represented apart from Severus Alexander and Maximinus Thrax). The hoard is important because it is the only one found in the Weald itself rather than on the periphery (though noting here the comparative lack of archaeological investigation in this region). As set out in Section 3.4, it has been examined in detail by Stuart-Hutcheson at the British Museum working with Sam Moorhead, she (2012) saying that:

> *"The High Weald hoard is similar to the famous Dorchester hoard as it contains a high proportion of earlier, less debased coinage such as that of the emperors Gordian and the Philips. However, the hoard is unusual as it closes with Postumus but is not composed highly of the extremely debased coins, such as the Bassaleg, Caerleon, Eastbourne and Selsey hoards. It is, furthermore, rare for such a large hoard to contain no examples of the smaller denomination denarii."*

A mercantile origin for many of the coins can be inferred by the wide geographic range of mints from where the coins originated. These include Rome, Antioch, Milan and Lyons. The hoard also contains a comparatively high number of rare coins, for example a radiate of Gordian III's wife Sabina Tranquillina and coins documenting the Secular Games of Phillip 1. These rare coins, together with the small number of highly debased radiates and the hoard's location near the Grade 3 site at Bardown with its known *Classis Britannica* association, may therefore indicate it had a high status origin with a state-association.

Analogy has also been considered in this research for the first time with regard to the nature of the Wealden occupation-period iron industry. For example, while there is comparatively little epigraphy outside the examples cited above from the Weald to link the region to a state/ *Classis Britannica*/ Imperial Estate association (even taking into account that stone inscriptions are generally less common in Britain than elsewhere in the Empire, with comparatively smaller assemblages of finds), other useful examples proliferate in other similarly-sized *metalla* across the Empire. Particularly useful are those which define mining or quarrying activity in a specific territory. Examples include inscriptions on boundary markers, funerary monuments, other types of stamped brick and tile, worked blocks of stone and graffiti. Hirt (2010, 48) notes that this is particularly valuable with regard to mining and quarrying enterprises because such activities, at least those on a large scale, took place in strictly defined territorial entities separate from the colonial, municipal and other official territories within a province. The two key words utilised in the epigraphy to show such official territorial boundaries are *prata* and *territorium*, both being used in association with military units or state activity. An example of the former is a boundary stone separating the *prata* of *legio IV* from the lands of the town of Segisamo in Spain, while for the latter one can look to a pronouncement by Severus Alexander (Emperor from AD 222 to 235) regarding a bath house built on the *territorium* of *legio II Augusta*. Le Bohec (2000, 220) explains that there is no agreed explanation concerning the differing uses of the two words, noting that theories range from the former being a sub-unit of the latter (and vice versa) to the former being used in an economic context and the latter for administrative or military purposes. The most relevant to state run industry appears to be *territorium*, a good example being the use of *territoria metallorum* on the famous bronze tablets found in the early 20th century at the Roman copper, gold and silver mines at Vipasca in Portugal, clearly identifying the area as an administrative district (Hirt, 2010, 48). The tablets also identify a *procurator metallorum* as the state official managing this enterprise, definitive evidence that the site was state-run, although it is not clear if the official managed all the mines and quarries of the province, some of them, or just this one (and by extrapolation whether *territorium,* or indeed *prata*, referred holistically to an Imperial estate, or to part of one). Whatever the breadth of his responsibilities, he would have reported directly to the provincial Procurator and thence to the Imperial *fiscus*. Hirt (2010, 106) also points to the large scale quarries at sites such as Mons Claudianus, Mons Porphyites, Tibarian and Mons Ophiates in Egypt, Dokimeoin in Asia Minor, Simitthus in North Africa and Karystos on Euboea as other key examples where epigraphy provides evidence of direct state involvement.

To bring the focus of these analogies back to the Weald, outside of the specific epigraphy already detailed, the key link (given the lack of a specific mention of *prata, territorium* or a bespoke *procurator metallorum*) is that of scale. It is very clear from the above examples that it was completely normal for *metalla* of this scale (for example Vipasca) to have some form of state-association, and with the eastern/ coastal Wealden iron industry we see industrial enterprises exploiting natural resources of just this size. In the case of Beauport Park alone (with its *Classis Britannica* tile and other epigraphy), this site produced some 30,000 tonnes of slag and waste, making it one of the largest such sites across the whole of Empire, and it is noteworthy that all three sites producing waste volumes of over 10,000m^3 are in the eastern/ coastal region (Beauport Park, Oaklands Park and Footlands Farm, see Section 3.1.4 above, Hodgkinson, 2009, 31). By way of comparison, as set out in Sections 3.2 and 3.5 above, Jackson is clear that the smaller (at least through to the mid-3rd century AD) iron-manufacturing operation at Ariconium in the Forest of Dean shows no evidence at all of a state/ military presence.

A final point newly considered here concerns the seeming lack of reinvestment of wealth (certainly in terms of conspicuous consumption outside of the limited amount of ubiquitous Samian ware detailed in Section 3.1.4) in the occupation-period Weald, despite a clearly thriving iron manufacturing industry which exported far and wide. Though some of the value may have been used to develop infrastructure (for example roads and port facilities), the lack of evidence more broadly of a desire to overtly display *Romanitas* is striking. Using Westhawk Farm as an example, Willis (2012, 434) says that this could point to the presence of absentee owners in the form of the state, he adding: *"It is possible that wealth, like the iron products, went elsewhere, reinforcing a suggestion that the community and its enterprises were subject to a controlling military and Imperial force."*

He adds this contrasts with the evident existence of community wealth at the site both before and after this evident period of state interest at Westhawk Farm.

As can be seen above there is a compelling body of data and opinion-based evidence to suggest a *Classis Britannica* interpretation for the management and more of the eastern/ coastal region of the Wealden iron industry during the occupation. Further, if one sets the above evidence against that set out for Imperial Estate interpretations in Section 2.2.4, the case in this regard is also strong, with positive answers for all five evidential questions set out there - lack of settlement (especially villas), unusual transport networks, unusual land use patterns (with little evidence of extensive agriculture excepting the exploitation of woodland for the iron industry), an association with other industry (that manufacturing tile and brick here, clearly I believe run by the *Classis Britannica*) and the presence of the Roman military in the form of the regional fleet.

Therefore, in my opinion, while the case is not 100% proven (for example through the location of epigraphy specifically mentioning the procurator metallorum), there is enough evidence for a very strong case to be made that the extractive industries exploiting natural resources in this eastern/ coastal Wealden *metalla* were being managed by the State through the services of the *Classis Britannica* as an industrial Imperial Estate.

5.1.4 The State Presence in the Medway Valley

As with the iron-industry in the occupation-period Weald, a number of commentators have argued that the state also played a major role in the upper Medway Valley ragstone quarrying industry. These include Marsden (1994, 83), Milne (2000, 131) and Pearson (2002a, 44), with the former going even further in suggesting an industrial Imperial Estate interpretation (1994, 83, see quote below). As is made clear above in Section 4.1.1 however, prior to the advent of this research the body of evidence used to support this interpretation has been far less defined than that for the Weald. To that end, as with Section 6.1.3 above, here I firstly set out the existing data and interpretations, before considering the new evidence gathered within this work.

In the first instance, the most prominent evidence cited to link the ragstone quarries of the upper Medway Valley with state-run *metalla* is the sheer scale of their output. The best example remains the original late 2nd/ early 3rd century AD 3.2km land wall circuit of London, detailed above in Section 5.2, which Hall and Merrifield (1986, 28) said comprised more than one million squared and dressed ragstone blocks. To this high profile use Marsden (1994, 84) also adds upper Medway Valley ragstone being used in the construction of the basilica, *forum*, at least three public baths, the Governor's palace and a wide variety of public buildings in London, with Bateman (2011, 31) adding the second phase of the amphitheatre. Marsden (1994, 83) is explicit in linking quarrying on this scale with the state being involved, saying: *"...it seems likely the (upper Medway Valley) area was part of an 'Imperial Estate' owned by the Emperor, for this would guarantee the output of the quarries over a long period on this scale."*

From the consideration of scale hinting at state involvement, the existing commentators next turn to the importance of maritime transport to the success of the ragstone quarrying industry for further evidence (this bringing the *Classis Britannica* into view). Roman London expert Merrifield (1965, 49), as detailed in Section 4.5, was blunt on the subject and we have hard data to support this view in the form of Marsden's enigmatic Blackfriars 1 vessel found on the bed of the River Thames in 1962 with a load of upper Medway Valley ragstone still in its hold. Milne (2000, 131) argues that this vessel, together with others of similar design found in the area of operations of the *Classis Britannica*, were specific to the regional fleet and therefore indicators of its involvement with the ragstone quarrying industry here.

The adherents to the state running this *metalla* next turn to non-maritime transport infrastructure as evidence, particularly the Rochester–Wealden Road linking the north Kent coast with the major iron working sites in the eastern/ coastal Weald (Marsden, 1994, 83). Margary (1967, 44) showed how closely this road tracked the upper Medway Valley with its quarries before heading south, it being utilised as set out in Section 2.5 and Section 3.3 more for administration than for the transport of regionally extracted natural resources which would as suggested above have principally used maritime routes. New data about this key regional routeway is revealed below as part of this research.

Analogy is also used by those arguing for a state-presence exploiting the ragstone resources of the upper Medway Valley. Russell (2013a, 41) says that one example would be the Mons Claudianus granodiorite mine in the eastern Egyptian desert, where epigraphic evidence provided by letters, passes and receipts shows at least 20 Centurions seconded there from the local legions and, topically for this research, the *Classis Alexandrina*. Another analogy closer to home would be the *Classis Germanica* (notably the only other provincial fleet to feature a centenary rank commander, see Section 2.6.1 above) with its quarrying activities along the Rhine and its tributaries (S. Elliott, 2014b, 50). Epigraphic evidence of this comes from numerous naval inscriptions in the Trass quarries on the left bank of the Brohol Valley, and from similar evidence that vexillations of this fleet quarried tufa for the Trajanic colonia Ulpia at Vetara. Meanwhile, more evidence of the British regional fleet carrying out quarrying activities comes from even closer to hand, with for example the inscription at Benwell fort on Hadrian's Wall which shows the regional fleet constructing the granary there (RIB1340, Breeze and Dobson, 2000, 66). Such responsibility for installations on the wall would have included quarrying the necessary stone as well as construction (Breeze and Dobson, 2000, 83). Further, on the doorstep of the upper Medway Valley ragstone industry is of course the Wealden iron industry referenced above in Section 5.1.3, which Marsden (1994, 83) argues was a direct analogy in terms of state-presence with the *metalla* of the upper Medway Valley.

Anecdote has also been used by existing commentators to support the state-presence in this region, specifically in the context of chronology. Milne (2000, 131) for example has pointed to the synergy between the ending in the mid-3rd century AD of industrial scale ragstone quarrying in the upper Medway Valley and the disappearance of the *Classis Britannica* (which he argues was the state representative facilitating the quarrying).

Before moving on to new data made available through this research, one question now needs to be addressed directly. This relates to the lack of *Classis Britannica*-stamped tile in the region, or indeed any other tiles featuring an official stamp (the furthest north coming from Cranbrook in mid-Kent, Brodribb, 1970, 1). As is detailed above in Section 3.6, a thriving tile and brick industry existed alongside the iron industry in the Weald. Both are directly associated with the *Classis Britannica* given the huge quantities of tiles featuring the regional navy's stamp known from the major iron working sites in the eastern/ coastal region, this tile being particularly important given their use there to help identify the regional fleet's presence. The lack of such tile in the Medway Valley is thus troubling, though I believe the answer *is* a simple one - there was no need. The lower Medway Valley had its own prolific early-occupation period tile industry before the *Classis Britannica* had officially come into being (the tile therefore not being stamped, either by the regional navy or indeed with any other official mark). These cream, yellow and white tiles are thought to have been made near the villa estate at Eccles downriver of modern Maidstone where the river was tidal then as now. This ensured their wide distribution in vast quantities down the lower Medway Valley and up the Thames Estuary as far as London. Thus by the time the *Classis Britannica* is first mentioned in the later 1st century AD and the Wealden tileries initiated production in the early 2nd century AD, local sourcing of tile in the Medway Valley was already mature and had been happily catering for regional needs since the beginning of the occupation. There was therefore no need for the *Classis Britannica* to step in.

Moving onto the new data made available here, in the first instance we can return to scale again, though this time in the context of the enormous ragstone quarries whose specific locations have been newly revealed for the first time in this research. To recap, these five quarries had an area of 61,600m^2 in the case of Allington, 54,600m^2 for Boughton Monchelsea, 356,400m^2 for Dean Street, 215,000m^2 for Quarry Wood and 35,830m^2 for Teston. They were clearly industrial (in the modern sense of the word) in scale, with that at Dean Street being comparable in size to the largest *metalla* across the entirety of the Roman Empire, for example at Rio Tinto (Jones, 1980, 148). It is here where analogy is useful again, given mining and quarrying operations on this scale were heavily supported by the state elsewhere in the Empire (Hirt, 2010, 106, Cleere and Crossley, 1995, 66, ***de la Bédoyère, 1992, 100 and*** Jones and Mattingly, 1990, 192), as fully detailed in Section 5.1.2 above.

Next, through the creation for the first time in this work of the distribution table showing the regional supply network for ragstone quarried in the upper Medway Valley, we have significantly added to our knowledge of the extensive maritime distribution system which played such an important role in the urbanisation and early fortification of the south and east of occupied Britain. Notable here are the parallels between the distribution of ragstone and Wealden Fabric 2 tiles, illustrating in detail the mature nature of this maritime system as detailed in Section 2.5. It can be argued that the systematic nature of this trading network up and down the east coast (as defined by Morris, 2010, 1, and Evans, 2013, 433, again see Section 2.5) speaks to the organisational presence of the *Classis Britannica* and is certainly a key indicator of the sophisticated levels of market integration on display here. Sticking with transport, research associated with this work has also located the Roman roadway from the Dean Street quarry to the Barming ford and East Farleigh villa. This route becomes particularly important if it can be proven to be a spur of the Rochester – Wealden road, it then being a very specific and direct link between the Wealden iron working sites with their state/ *Classis Britannica*/ Imperial Estate provenance as argued in Section 5.1.3, and the upper Medway Valley ragstone quarries.

This research has also, as set out in Section 5.3 above, specifically linked for the first time the elite settlements in the upper Medway Valley (in the form of the villa estates ranging from Allington through to Teston) with their associated quarries, making the case that one interpretation would see them as the country residences of those actually managing the quarrying activity through to the mid-3rd century AD (though see discussion below). If they were such, they would have formed an elite enclave where the owners competed with neighbouring friends to display wealth and culture via architectural form, exhibiting conspicuous consumption at the same time. This linking of the villas and quarries has included the identification of Building 5 at the East Farleigh villa site as a temple, indicating the significance of this particular settlement, and the possible bridge under investigation at Tovil, detailed in Section 4.1.4, which would link for the first time the Dean Street quarry on the south bank of the River Medway with the Florence Road/ Bower Lane villa estate to the north. Other evidence of settlement activity near to the quarrying sites has also been revealed, for example the 'barrack' crop mark and LIDAR image close to the Boughton and Dean Street quarries on the site of the *'Roman foundations and coins found 1860'* as detailed in Section 4.3.3 above.

However, the major breakthrough however in terms of demonstrating a specific link between the occupation-period upper Medway Valley ragstone quarries and the state could come through further investigation of the 'Medway Stones' site in the River Medway between Tovil and East Farleigh (Elliott, 2014c, 11). Fully detailed in Section 4.1.4, the author believes this is a candidate occupation-period wreck carrying a cargo-load of worked stone (the 'Medway Stones' themselves being those dredged to the surface to date). A full-scale underwater investigation is planned. As detailed above, Marsden's Blackfriar's 1 vessel with its load of upper Medway Valley ragstone is already being used by existing commentators to suggest a link between industry here and the state (Marsden, 1994, 83). A second such wreck, found *in situ* adjacent to a quay in the River Medway itself, would significantly advance such a hypothesis, especially if the design of the vessel transpires to replicate that of Marsden's Blackfriars 1 ship (Milne, 2000, 131).

Most recently, data from the location of the Gallants Lane iron-working site between the Dean Street and Quarry Wood quarries has also significantly advanced our understanding of the wider economic landscape within which the ragstone quarries sat. Confirmation of the attribution of this site after further investigation would for the first time directly link iron working amid the ragstone quarries in the upper Medway Valley with that in the Weald (as I argue above in Section 5.1.3, with the latter's likely state/ *Classis Britannica*/ Imperial Estate provenance). It would also illustrate the importance of the nearby Rochester – Wealden road in connecting the two *metella*, and could be used to help build the case that they are one and the same state-run *metalla* enterprise.

The final piece of new data generated by this research relevant to the consideration of a state-role in the upper Medway Valley ragstone quarrying industry is the recreation for the first time of a typical journey for a Blackfriars 1-style vessel with its load of extracted stone to London and back. This, and a simple review of a map showing the inherently difficult 254km round trip, can be used I believe to argue that a state presence was likely to ensure its smooth running in order to provide the huge quantities of stone that the demand required (this being a core theory advanced by Marsden, 1994, 83). The synergy between the disappearance of the *Classis Britannica*, the likely state-actor in the above hypothesis, and the ending of industrial-scale ragstone quarrying in the Upper Medway Valley, is also particularly telling in this regard (as with the eastern/ coastal Weald and its iron manufacturing industry).

In the above discussion I have considered existing and new data to determine whether a case can be made that the state, through the *Classis Britannica* as its military representative, controlled the ragstone quarrying industry in the upper Medway Valley until the demise of both in the mid-3rd century AD. When one considers the sheer scale involved, both in terms of the of the quantity of ragstone quarried and used, and the parallel scale of the maritime commitment to facilitate this industry, together with the facilitating administrative land links to the eastern/ coastal Weald with its known military association, the emerging wider industrial landscape with at least one potential iron working site, the emerging link between elite and other settlement with the ragstone quarries, and the wealth of supporting analogy (not least the proximity to the likely *Classis Britannica* controlled *metalla* in the Weald as mentioned) and anecdote, then a case does begin to emerge. Certainly this industry was part of the Imperial economy rather than the provincial economy, with again the sheer evident scale pushing towards the Imperial Estate interpretation as believed by Marsden (1995, 84).

As with the Weald though, where to be candid the evidence for a link between the military and iron manufacturing in the eastern/ coastal region is stronger, there is to date no definitive evidence, for example in the form of undisputable epigraphy. It so happens that Kentish ragstone was not a stone favoured for inscriptions and the main floruit of its employment, particularly in London and elsewhere (and thus its quarrying) occured at a time when inscriptions in stone were becoming less common. Also quarries and villas are not normally locales with stone inscriptions, which is another reason why epigraphic indications have not to date been forthcoming. The question therefore is how strong the case is today for a military presence running an Imperial Estate in the occupation-period Medway Valley metalla, based on the available data. Here I again turn to the model set out in Section 2.2.4 to help interpret a geographic economic entity as an Imperial Estate, with its five paradigms, and what emerges is a largely positive though confusing picture. In terms of the positive, the region does feature unusual transport networks (for example the riverine hydraulic infrastructure which would have been necessary to allow the River Medway to be used above the tidal reach on this industrial scale), unusual land use patterns (such a concentration of enormous quarries is most uncommon during the occupation), an association with other industry (the newly found iron working site for example) and the (likely, based on the scale of the required maritime commitment and proximity to the Weald) presence of the Roman military in the form of the regional fleet. What counts against the Imperial Estate interpretation in terms of the five paradigms though is the lack of unusual settlement patterns given the fact that we have a dynamic range of villa estates along the banks of the upper Medway Valley, together with other settlement (unlike in the Weald where the absence of such amid the large iron working sites is notable). To counter this, I do make the case above that these villas and settlements may have been associated with the quarries themselves, and Mattingly (2006, 371, see discussion on the Roman economy in Section 2.2.4) does argue there is no good reason that villas would not feature in an Industrial Estates landscape (though common senses here indicates this would have been more likely with regard to agricultural Imperial Estates).

The honest answer here is that the evidence for the ragstone quarrying industry in the upper Medway Valley being a military-run Imperial Estate is less clear cut that that for the iron manufacturing industry in the Weald, excepting perhaps its scale and the enormous maritime commitment to ensure its success. I think one can certainly say that it was part of the Imperial economy and under state control (the exploitation of natural resources as part of the Imperial economy being a matter of state control, Mattingly, 2006, 494), and further that it was under the direct control of the Procurator through his *procurator metallorum*, if not as an Imperial Estate (run by the military or others) then using tightly controlled contractors at the very least (likely based in London in this interpretation, with a number of the villa estates then occupied by their representatives, for example local elites employed in this regard or *vilici*). Over and above that though, based on existing data and interpretations, I am reduced to expressing an informed opinion. In that regard I think a strong case can be made that the *Classis Britannica* was involved in some way in this enormous industrial enterprise, if only to facilitate the required transport network, though with the Imperial Estate interpretation being at present unprovable.

5.1.5 Synthesis

Above I have considered the existing and new data and interpretations available to support the hypothesis that the state, often in the form of the *Classis Britannica*, supported the exploitation of natural resources in Kent and the South East during the Roman occupation. In this regard the research has specifically focused on two of the principal areas of such resource extraction, namely the iron industry in the Weald and the ragstone quarrying industry in the Medway Valley. Broadly, the investigation has shown that the scale of industrial activity at both, together with other supporting evidence, is such that the case for a significant state presence can be made, even though it is definitively unprovable to date.

With regard to the Weald and upper Medway Valley, a point for further consideration is whether the two industrial zones were linked as one *metalla* under the same *procurator metallorum* tasked with ensuring they contributed to their maximum potential for the Imperial *fiscus* (given their strategic and financial importance). Again I believe that a case can be made in this regard, though this also remains unprovable based on existing data (especially given the case for an Imperial Estate interpretation for the iron industry in the Weald is stronger than that for the ragstone quarrying industry in the upper Medway Valley).

5.2 Change and Continuity in the Extractive Industries of the Occupied South East

As set out in the introduction to this chapter, change in the exploitation of natural resources by the extractive industries is evident in Kent and the South East in a pronounced way from the mid-3rd century AD. This development is well illustrated by the data evidenced in Chapters 3 and 4, whether relating to industry large and small, or to settlement. Specific evidence for this pronounced change in the economic outlook of the region is many and varied. We have of course the arguments regarding the disappearance of the industrial enterprises in the eastern/ coastal Weald and upper Medway Valley. On the one-hand the monumental manufactories producing iron for Roman-occupied Britain and export simply ceased to exist, while on the other the large-scale use of ragstone across the region as a building stone dropped in scale significantly (this despite the initiation of the fortification process around the Saxon Shore at this time). These developments are discussed in full in Sections 5.2.2 and 5.2.3 below using existing and then new data and interpretations to provide fresh insight into the dramatic events described.

These discussions follow a brief review of the transition in the region from the LIA to the occupation to provide context, with Section 5.2 then concluding with a synthesis of the presented data to discuss the possible reasons behind these dramatic changes in the mid-3rd century AD.

5.2.1 The Late Iron Age/ Occupation Transition

Kent and the South East of Britain during the LIA were already beginning to display cultural change as the impact of the Roman conquest of Gaul under Caesar (and indeed his two incursions into Britain in 55 and 54 BC) began to impact political, economic and social structures. Champion (2007, 132) says this first manifested itself in a coalescence of power around a few individuals in the mid-to-later first century BC, as evidenced by a new and richer stratum of burials in the region. As the century progressed he argues this regularised contact with Rome led to more evident changes in material culture, for example the use of Roman prototypes for coinage which he says was evidence of closer contact between the region's ruling families and Rome. This engagement between the elites of the South East of Britain and the Empire also facilitated the latter's growing awareness of the natural resources available in the region for exploitation, a key factor in the rapid growth in the *metalla* here shortly after the occupation began.

Turning firstly to the Weald, as set out in Section 3.1.2 above, iron production began here as early as the 8th century BC, with Hodgkinson (2008, 28) detailing that up to 23 bloomery sites may have been active by the time the LIA drew to a close. Given Caesar's reference to '*iron in the maritime*' (The Conquest of Gaul, V.135) and Strabo's comment about iron being exported from Britain (The Geography, IV.5), the potential for the large-scale exploitation of the siderite and timber resources here was clearly well known on the Continent. It is in this context that we should therefore view the almost immediate growth in iron manufacturing in the Weald shortly after the AD 43 Claudian invasion, catering for the demand created by the military expansion to the west and north (Bray, 2010, 175). The change here from the LIA to the occupation was thus comparatively smooth, with the larger sites in the eastern/ coastal region growing to operate at full capacity by the end of the century.

Change in the upper Medway Valley following the occupation was far more dramatic however. Here there is no evidence at all of significant industrial activity prior to the conquest (the nearest being the limited extraction of Ightam Stone for use as a primitive revetment over 100 year earlier at the Oldbury *oppida* in north west Kent, see Section 2.1), with the advent of the industrial-scale ragstone quarrying which was to define this region so clearly during the first half of the occupation directly associated with the arrival of Rome (S. Elliott, 2014a, 252). Clearly given the rapid growth of this industry from a zero-base, with an integrated transport network to complement the actual resource extraction, there was again detailed foreknowledge of the natural resources available for exploitation. As Houliston (1999, 163), Rowsome (1999, 262) and Pearson (2002a, 82) detail, the early appearance of finely-worked ragstone from the upper Medway Valley in the Claudian temple at Colchester (70km away as the crow flies) and in London (50km away, for example at the Huggin Hill bath house) are definitive evidence of this.

In both regions of study regarding the exploitation of natural resources we can therefore see change initiated by the LIA/ occupation transition, less dramatic in the first than in the latter.

5.2.2 Change and Continuity in the Weald

As detailed in Section 3.4, the iron manufacturing industry in the Weald flourished following the beginning of the Roman occupation, building on its humble LIA beginnings (see Sections 3.1.2 and 5.2.1) to become one of the largest industrial enterprises across the entirety of the Empire (Cleere and Crossley, 1995, 81). This research, building on the work of Harrington and Welch (2014, 109) and others, has shown definitively that this industry was split into two distinct regions of iron-manufacturing activity, a central one where the industrial sites were (for the most part) smaller and catered for regional demand, and an eastern/ coastal one where activity was on a far more epic scale and catered for national and indeed Continental demand. I further suggest above in Section 5.1.3 that a strong case can be made that the state, in the form of the *Classis Britannica*, played a major role in running the *metalla* in this latter region, and it is here where we now look to see a major and negative change taking place from the mid-3rd century AD.

Existing data supporting this interpretation comes from the research of commentators including Cleere (1977, 18), Brodribb *et al* (1988, 232), Cunliffe (1988b, 86), Jones and Mattingly (1990, 193), Lyne (1994, 545), Booth (2001, 3) and Hodgkinson (2012, 1). They have shown that by the mid-3rd century AD iron manufacture at the major complexes in the eastern/ coastal region including Bardown, Crowhurst Park and Beauport Park had ceased, and was also finishing at their associated satellite sites. Further, iron production at the more peripheral sites in the region including Westhawk Farm was also ending. Meanwhile the maritime infrastructure built around Romney Marsh to support industry here also disappeared at this time, while an economic change in the nature of pottery manufacturing in the region in the mid-3rd century AD (from wheel-worked high quality pottery to handmade ware) also hints at major change. In the latter regard Lyne (1994, 545) links this directly with the decline of the iron industry, he suggesting the skilled pottery workers dispersed elsewhere in occupied Britain after regional demand for their wears disappeared. Hodgkinson (2008, 34) says that only at Footlands Farm is any iron working visible in the eastern/ coastal region into the 4th century AD, while Cleere and Crossley (1995, 81) highlight the scale of this decline across the whole of the Weald, saying that annual iron production figures between AD 350 and AD 400 were only 50 tonnes (and then exclusively in the central region), down from a peak of 750 tonnes a year between AD 150 and AD 250.

In terms of new data to support this interpretation of a dramatic decline we can turn again to the High Weald coin hoard (Stuart-Hutcheson, 2012, though noting this is only one hoard). Of particular interest given its location at a Bardown satellite iron-working site, and the view as detailed in Section 5.1.3 that it represented some kind of official presence, the latest coin here dates to AD 268. As such the hoard is illustrative of the abandonment of this region as the final sites closed one by one, leaving a discarded industrial landscape which would remain lost to memory until its rediscovery by antiquarians many centuries later.

5.2.3 Change and Continuity in the Medway Valley

As hinted above in Section 5.2, the change in industrial activity in the upper Medway Valley from the mid-3rd century AD was dramatic and rapid, with both the existing and new data showing how it impacted not only the monumental-scale ragstone quarries but also the associated elite settlements linked to the industrial sites in this research for the first time. Those who highlight this change using existing data include Merrifield (1965, 48), Blagg (1980, 5), Williams (1993, 31), Milne (2000, 131), Pearson (Pearson, 2002a, 82, and 2006, 30) and Sheldon (2011, 230). The best evidence in this regard is the apparent ending of the use of ragstone for the construction of buildings both public and private at this time, its last possible use on any scale being in the land walls of Canterbury constructed around AD 270 (and even here it was limited to use as footings and lower facings). Prior to this upper Medway Valley ragstone was one of the principal building stones of choice throughout the whole region, from London in the west, Colchester to the north and Dover in the east. Afterwards it was replaced by a variety of alternatives, often featuring the re-use of existing building materials. A good example of the latter can be found with the walls of Roman London, the late 2nd/ early 3rd century AD land circuit being constructed of thousands of skilfully worked blocks of upper Medway Valley ragstone but the late 3rd century AD river wall and associated bastions being constructed of roughly reworked materials re-used from demolished public buildings and mausoleums. Similarly the late 3rd century AD 'Carausian/ Allectan' palace excavated at Peter's Hill in London (or temple complex, Bradley and Butler, 2008, 9) was constructed using re-worked stone (Williams, 1993, 31), as were significant parts of a number of Saxon Shore forts, with Fields (2006, 21) highlighting Richborough and Lympne as examples.

Meanwhile new data considered in this work in the context of change here relates to settlement, with disruption evident at a number of elite sites at this time, this being detailed by Detsicas (1967, 173), Davies (1982, 137), Houliston (1999, 71), Dawkes (2009, 1) and Daniels (2015, 6). Examples include Snodland on the Roman-period tidal reach (Kaye 2015b, 232) where the villa was significantly modified at this time. Here, the principal bath house was demolished and replaced by a larger aisled building, with two timber buildings constructed at the same time to the north and south. Notably the surrounding field system was also reorganised. Meanwhile at Allington the Tetrican coin hoard found in association with the possible villa settlement, quarry and burials is also indicative of some kind of regional disruption (though again noting that 3rd century AD hoarding is also often associated with coinage devaluation), while at the Mount in Maidstone (with its association with the Boughton ragstone quarry) significant rebuilding also took place. Further upriver at the East Farleigh villa site, linked directly by road with the nearby monumental Dean Street quarry, change is also evident, with the main structure rebuilt on a new alignment. Even at the villa sites downriver of the tidal reach change is apparent, for example at the large elite settlement at Eccles where another reconstruction of the villa on a new alignment took place.

The reasons behind these changes to the industrial sites and associated settlement in the upper Medway Valley are now discussed below in a synthesised review of such developments across the whole region.

5.2.4 Synthesis

Given the weight of the evidence presented above in Sections 5.2.2 and 5.2.3 it is very clear that great change did indeed take place in terms of the exploitation of natural resources (and associated settlement) in the Weald and the upper Medway Valley in the mid 3rd century AD. There are numerous, often linked, reasons why this change may have taken place and here they are considered at both an Empire-wide and regional level. As each potential reason is considered, the data presented above is tested to determine its validity with regard to each hypothesis.

Looking first at the big picture, in the 3rd century AD the Empire was hit by a succession of major issues so severe that by the time it emerged into the comparative stability of the early 4th century AD it was a different political, economic and social entity. These issues included:

- Conflict, with significant invasions of its territory from opponents whose contact with Empire had precipitated a level of critical mass previously absent (Heather, 2013, xv).
- Frequent civil war following the murder of Emperor Severus Alexander in AD 235, with a troubled Imperial succession continually being a catalyst for internal conflict (Whitby, 2002, 27).
- Pestilence, for example the '*Plague of Cyprian*' which afflicted the Empire from around AD 250 until after AD 270, thought to be a virulent form of smallpox which was claiming 5,000 lives a day in Rome at its height (Stathakopoulos, 2007, 95).

- Economic depression as Imperial expansion slowed and ceased (Cornell, 1993, 168).

Together, these issues have given rise to the description of this period of Roman history (particularly from the death of Severus Alexander to the accession of Diocletian in AD 284) as the 'Crisis of the 3rd Century'. The impact of this crisis was clearly felt in occupied Britain (it being around this time that we see the fortification of many towns and the advent of the Saxon Shore forts in the South East), though perhaps to a lesser extent than on the Continent given the comparative lack of severe barbarian incursions (Halsall, 2013, 92). In terms of Kent and the South East however, given its cultural alignment with northern Gaul (Blanning, 2014, 484), our region of study was clearly more exposed to the convulsions taking place across the English Channel than elsewhere in Britain.

The region may in fact have been impacted both negatively and positively by the tribulations taking place on the Continent. In the former context this is self-evident given that a political and economic crisis shaking the Empire in its entirety would be felt in all of its provinces to a greater or lesser extent, with Kent and the South East in a British context being particularly exposed. Using the *metalla* of the upper Medway Valley as an example, at first the building of the urban wall circuits across the South East (though not the Saxon Shore forts) drove a late boom in the extraction of ragstone. This rapidly tailed off however and it is from this time that we begin to see evidence of industrial decline, with for example the virtual disappearance of ragstone as a regional construction material of choice. Pearson (2006, 30) contextualises this in arguing that the era of major public building in the urban environment in Britain finished around this time (see Section 5.2.4 above), he linking this development to the ending of the period of Imperial expansion.

Similarly, looking at rural developments, Blanning (2014, 480) has shown how the building of new elite country residences in the form of villas in Kent and the South East had significantly declined by this time, again impacting the demand for ragstone. In this context therefore, one could argue that by the mid-3rd century AD there was simply not enough demand to justify *metalla* so large that the state had to run them (as discussed regarding the Weald in Section 5.1.3 and upper Medway Valley at Section 5.1.4), with localised quarrying on a much smaller scale catering for the lesser demand (Everitt, 1986, 51).

Such economic disruption as events unfolded on the Continent is also evident in other forms of data, with Evans (2013, 433) arguing that by the middle of the 3rd century AD many of the elite continental imports coming into London via the emporia of eastern Kent as part of the east coast trade system (and indeed into Poole Harbour through his west coast trade system) had ceased. Further, he cites ceramic data to indicate that wine imports by this time were far less than at their height in the 2nd century AD, with olive oil also being imported at only a fraction of previous levels. The political impact of the turmoil on the Continent is also evident in the region, with Stuart-Hutcheson (2012) highlighting the 'massive' increase of coin hoarding in the mid-3rd century AD, for example those found in the High Weald and Allington (see Sections 5.2.2 and 5.2.3 above).

It is the positive impact however which may also help explain the change across Kent and the South East in the mid-3rd century AD, particularly with regard to the balance of the regional economy. Large-scale industry certainly dominated this economy in the early occupation through to the mid-3rd century AD (S. Elliott, 2014b, 49). After this time however it declined dramatically, with other areas of the country picking up the mantle of industrial heartland, for example in the case of iron manufacturing the Forest of Dean (Jackson, 2012, 169) and East Midlands (Mattingly, 2006, 509). In place of industry in Kent and the South East we see from this time the growing importance of agriculture, driven by a new demand from the Continent given the Imperial need to feed a north west of Empire increasingly disrupted by the 'crisis' detailed above. This again reflects the region's proximity to, and alignment with, northern Gaul (Blanning, 2014, 484). In Kent and the South East we may have direct evidential data of this switch to the dominance of agriculture, with the expansion of the granary at the Horton Kirby villa site in the Darent Valley to a huge size in the mid-3rd century AD, and the construction of the similarly large granary at Lullingstone towards the end of the century (Philp, 1972, Perring, 1991, 119, and Blanning 2014, 298).

Turning to regionally-specific events and issues which help explain the changes taking place across Kent and the South East in the mid-3rd century AD, and again referring to the data outlined in Sections 5.2.2 and 5.2.3, in the first instance we can consider the problematic disappearance from the archaeological and historical record of the *Classis Britannica*. This is important given the argued-for links between the regional navy and the *metalla* of the eastern/coastal Weald and upper Medway Valley.

Broadly the same issues seem to be in play here as those which affected every other manifestation of the Empire in the mid-3rd century AD. Certainly the 'crisis' accelerated military reforms already initiated by Septimius Severus at the end of the 2nd century AD, such that by the time the further military reforms of Diocletian in the late 3rd century AD and those of Constantine in the 4th century AD were complete, the shape of the Roman military was very different from that of the Empire at its height in the mid-2nd century AD. Within this narrative of change it remains a possibility that the *Classis Britannica* was disbanded during the rapid turnover of Emperors in the middle of the 3rd century AD at the height of the 'crisis', perhaps finding itself on the wrong side of a usurpation attempt. The context may have been the 20-year power struggle between the military and Senate after the death of Severus Alexander, though events during the break-away Gallic Empire from AD 260 to AD 274 may provide a more likely setting for its demise. Other candidates include at the time of the separation of the original province into two by Septimius Severus or his son Caracalla in the early 3rd century AD (though this seems too early given the last reference we have to the regional navy is dated AD 249), and at the time of the Carausian Revolt from AD 286 to AD 296, though this seems to the author a little too late.

Whenever it happened however, we do definitively know that the *Classis Britannica* disappeared. As to whether this would have played a role in the changes taking place in our region of study, certainly the Procurator would have felt the impact of the disappearance of the body actually running his key *metalla* in Kent and the South East if that were the case as discussed in Section 5.1.4. Put simply, given the evident scale of the *metalla* here, there may have been no alternative agent for the state to use under this hypothesis (S. Elliott, 2014b, 49). Allen and Fulford (1999, 181) relect this view in arguing that as a result of the regional fleet's decline much wider-ranging systems of state-control and transport broke down, while Hodgkinson specifically references the drop in iron production following its demise (2008, 34). Under this hypothesis the disappearance of the *Classis Britannica* could therefore have played a very specific, regional role in the changes taking place in Kent and the South East in the mid-3rd century AD.

P. Elliott (2014, 23) also highlights a more insidious means by which changes taking place in the Empire and which impacted the *Classis Britannica* might also have had a knock on effect on industry and the broader economy in Kent and the South East, thus contributing to regional change. This is with regard to the big increases in pay for the military in the early part of the 3rd century AD instituted by Septimius Severus (from 300 to 450 *denarii* per year for a legionary, way ahead of inflation) and later his son Caracalla (he increasing pay further such that it was twice what it had been under Commodus in the late 2nd century AD). Such increases, later exacerbated as the 'crisis' took hold with each Imperial contender trying buy favour with the military, placed an increasingly heavy burden on the *fiscus*, with the currency ultimately being debased. This would certainly have had an economic impact on the profitability of the iron manufacturing sites and ragstone quarries if they were being state-run through the *Classis Britannica* as argued in Sections 5.1.3 and 5.1.4. Indeed it may be no coincidence it is in the middle of the 3rd century AD, when the economic impact of the increase in pay for the military was beginning to be felt in detail, that we see the decline of the Wealden iron industry and the upper Medway Valley ragstone industry. Therefore, in this context, perhaps the Procurator simply had to stop using the regional navy given the cost of doing so was undermining the regional economy. With no other such organisation to turn to this resulted in industrial decline and thus regional change.

Septimius Severus may also have played a more direct role in the decline specifically of the iron industry in the Weald, thus initiating regional change there in the mid-3rd century AD. Southern (2013, 251) and Evans (2013, 433) believe the campaigns of the Emperor and his sons in Scotland in the early 3rd century AD, while not actually 'conquering' the northernmost tip of the province, were so successful given their brutality that comparative peace reigned across the northern frontier for four generations. In this context the resulting lack of demand by the military in the north for iron produced in the Weald (noting that the Severan army for the Scottish campaigns numbered 50,000 men or more, Moorhead and Stuttard, 2012, 162) would clearly have had a major impact on the long-term prospects of the industry.

Other regionally-specific factors leading to mid-3rd century AD change in occupied Kent and the South East included the changing security environment. The early 3rd century AD saw the beginning of the construction of the Saxon Shore forts (Fields, 2006, 4), while as mentioned above fortification also took place at key urban sites. This is clear evidence of the emergence of an external threat existential enough to precipitate this significant commitment in terms of manpower and capital, not only in their building but subsequent maintenance and manning (Pearson, 2002, 39). In this regard Harrington and Welch (2014, 109) and others argue that one of the reasons for the decline of the eastern/ coastal Wealden iron industry was the exposure of the key sites to endemic raiding from the north European coast. Given the importance of maritime transport to the success of ragstone quarrying in the upper Medway Valley, a similar case can also be made for industry here.

Meanwhile, as detailed in Section 3.4, Cleere and Crossley (1995, 81) also argue that one of the causes of regional change specific to the eastern/ coastal Weald iron industry in the mid-3rd century was the silting up of the region's rivers which had provided the crucial access to the coast. Rudling (2013) adds here over-exploitation of the Wealden forests for timber as a further factor.

Taking all of the above into account, it is very clear that Kent and the South East suffered its own bespoke crisis in the middle of the 3rd century AD which changed the region significantly, such that as the 4th century AD approached the local economy and settlement patterns were significantly different from those at the height of industrial operations earlier in the occupation (S. Elliott, 2014b, 49). The reasons behind this were clearly complicated, with at a macro level the region certainly being impacted by the political instability on the Continent and elsewhere in the Empire, especially given its alignment with northern Gaul (Blanning, 2014, 484). Meanwhile with regard to regional causes we have the demise of *Classis Britannica* as a candidate, which to complicate matters may itself have succumbed to the same tribulations on the Continent as I discuss above in the broader sense led to industrial decline in the region. In this sense, if the regional navy's disappearance was a factor, it may have been the one of the many vectors for economic change generated by developments on the Continent rather than a separate regional issue.

Figure 48: The fertile Medway Valley at harvest time in East Farleigh (villa site centre left behind boundary hedge), arable farming and fruit orchards ready for harvesting. A useful analogy for the post-*metalla* industrial landscape of the later occupation. Simon Elliott.

Such is the level of complication when trying to use existing and new data to take a view about this Kentish occupation-period mid-life crisis. While the reality is that we will never know definitively the exact reasons why dramatic change took place here in the mid-3rd century AD, a combination of factors seems the most likely cause. The traumatic events on the Continent would certainly have played their part as the elites looked increasingly to agriculture (see Figure 48) to make up for an already present lack of demand for the exploited natural resources of the region, the latter phenomenon occurring for different reasons in the Weald and upper Medway Valley:

- For the iron industry in the eastern/ coastal Weald, the comparative peace on the northern frontiers following the Severan campaigns would certainly have impacted on the requirement for manufactured iron (with demand having peaked in the first 15 years of the 3rd century AD, Bray, 2010, 175).
- For the ragstone quarries of the upper Medway Valley, the ending of Imperial expansion clearly impacted demand for the stone-built public buildings of the first two centuries of the occupation. Demand for ragstone would then have peaked briefly at the beginning of the 3rd century with the first wave of regional fortifications, before a final and rapid dropping off.

For both, and along the east Kent coast, matters were exacerbated by the political turmoil across the English Channel, this also contributing to the demise of the *Classis Britannica*. If one adds to these other regional factors, for example the increasing risk of predation from the newly evident maritime threat from north-western Europe (especially if the regional navy was increasingly absent), then a picture comes into sharp focus of a region in dramatic transition. The occupation-period Kent and the South East that emerged on the other side of this change was very different, certainly featuring areas which thrived given the Continental demand for agricultural produce, but with a bleaker, even more militaristic feel, and with large areas such as the Weald largely abandoned.

6

Conclusion

This research has shown for the first time that in Kent and the South East during the Roman occupation there was a previously unknown depth to an industrial landscape which featured some of the largest extractive industrial enterprises exploiting natural resources in the whole province (and indeed Empire). They sat within an integrated system of transport (indicating a sophisticated level of market integration), supporting industry and settlement, and underwent profound change in the mid-3rd century AD.

Given the importance of this industrial exploitation, the research has provided a unique prism through which to track this change in a political, economic and social context across the region as it experienced its chronological journey as part of the Roman Empire. In that regard we have witnessed the LIA/ occupation transition as the Wealden iron industry grew exponentially and the ragstone quarries of the upper Medway Valley emerged to take advantage of a previously un-exploited high quality natural resource. We have also seen how these industries then suffered within the context of the 'Crisis of the 3rd Century', but in a particularly British and indeed Kentish and south eastern context, after which the region began a slow decline before ultimately experiencing a grim end of occupation.

I believe that the new data and interpretations made available through this research are especially relevant in an international context for those studying the Roman Empire, and Britain's experience of it, in three specific ways:

- The revelations regarding the scale of the extractive industries exploiting the natural resources, especially in the eastern/ coastal Weald and the upper Medway Valley. This forensic research, presented in an holistic context which has taken into account other industry and associated settlement, has enabled for the first time the iron manufacturing and ragstone quarrying *metalla* of Kent and the South East to be considered in the same context as their sister industries elsewhere in the Empire (for example the enormous silver, copper and lead mines at Rio Tinto in south western Spain, the gold, silver and copper mines at Vipasca in Portugal and the marble quarries at Dokimeion in Turkey).
- The discussion in Chapter 5, based on the existing and new data and interpretations set out in the two preceding regional studies, concerning the role played by the state in the regional *metalla*, and particularly the activities in that regard of the *Classis Britannica.*
- The examination in the two regional studies of specific aspects of industrial activity and the spotlight this has shone on the economic framework within which they played their part. A strong example would be the recreation of the journey of the load of ragstone on a Blackfriars 1-style vessel from the upper Medway Valley quarries to London, showing for the first time the complexities of such a journey (which would have taken place countless times during the occupation, as the principal city of the province was built and later fortified).

This work also provides insight into the individual experiences of the occupation, with agency writ large in the form of personal choice, opportunity or (at least in the context of the industrial workforce) coercion. The focus on regional industry has facilitated an understanding of this experience at all levels of society, whether with the *procurator metallorum* and his staff at the upper end ensuring the *metalla* contributed to their maximum potential for the Imperial *fiscus*, the *officinae* and *caesurae* in the mines and quarries facilitating their actual operation, the crews of the *Classis Britannica* vessels and privately owned merchantmen transporting the extracted natural resources around the coasts of Britain or across the English Channel and North Sea, or at the lower end the unskilled slaves and indentured workers damned to work literally at the rock face of the *metalla*. This phenomenological insight is also of value at an international level for those studying the Roman Empire (and classical industry more broadly) as it provides a comparator for the similar experiences of those living and working elsewhere in the world of Rome (noting how valuable I myself have found analogy in my own research).

As will be clear to those having read the book, the findings as presented here are not the end of the research given that there are many threads of investigation still to be completed. One could in fact say that completing the work is actually the beginning of the next phase of research, which I see as ongoing. As the reader may gather, one of my own principal ambitions is to use the research here as a springboard to initiate a fully holistic investigation of the upper Medway Valley during the occupation such that it is ultimately as well-known as Dr Brian Philp's east Kent coast, Dr Paul Wilkinson's north Kent coast and Henry Cleere and Jeremy Hodgkinson's Weald.

To conclude, all of the subjects and issues debated above in the two regional studies and the discussion point to Kent and the South East of Britain being a place of significance and difference in Roman occupied Britain and indeed the Empire. This is particularly the case with regard to industry and the associated role of the state. It had a distinctly utilitarian feel, though in parts and at certain times it displayed real political and economic (in an industrial context) grandeur, offering the ambitious elites arriving in the province or those born locally the prospect of fame and fortune as they headed north and west to campaign with the military or remain in the region to run the *metalla*. From the heady heights of the late 2nd and early 3rd centuries AD however the region experienced a steep decline, this increasing dramatically in the later 4th century AD when, at a time when other parts of the *diocese* were experiencing a late boom, Kent and the South East found itself increasingly cut adrift from its Continental life blood as the Western Empire began its collapse. By this time the booming *metalla* of previous centuries, and the life stories of all of those who had worked in them, were lost to living memory, awaiting rediscovery by antiquarians centuries hence and ultimately coming to life in this research in a holistic way for the first time.

Bibliography

Ancient Sources

Decimus Magnus Ausonius, *The Mosella* . 1933, Blakeney, E.H., London: Unknown Binding).

Julius Caesar, *The Conquest of Gaul*. 1951. Handford, S.A., London: Penguin.

Marcus Cato, *De Agri Cultura*. 1934. Ash, H.B. and Hooper, W.D., Harvard: Loeb Classical Library.

Cassius Dio, *Roman History* .1925. Cary, E., Harvard: Loeb Classical Library.

Herodian, *History of the Roman Empire*. 1989. Whitakker, C.R., Harvard, Loeb Classical Library.

Quintus Horatius Flaccus (Horace), *The Complete 'Odes' and 'Epodes'*. 2008. West, D., Oxford: Oxford Paperbacks.

Pliny the Elder, *Natural History*. 1940. Rackham, H., Harvard: Harvard University Press.

Pliny the Younger, *Epistularum Libri Decum*. 1963, Mynors, R.A.B., Oxford: Oxford Classical Texts - Clarendon Press.

Strabo, *The Geography*. 2014, Roler, D.W., Cambridge: Cambridge University Press.

Cornelius Tacitus, *The Agricola*. 1970, Mattingly, H., London: Penguin.

Marcus Terentius Varro, *Rerum Rusticarum*. 1932, Ash, H.B. and Hooper, W.D., Harvard: Loeb Classical Library, 1932.

Modern Sources

Ackroyd, P. 2012. *London Under.* London: Chatto and Windus.

Adler, S. D. 2013. *CL:BR. The Classis Britannica in Kent and Sussex*. Raleigh: lulu.com.

Aldridge, N. 1995. Three Notes on Ironworking Sites in Kent. *Wealden Iron – Bulletin of the Wealden Iron Research Group*, V.2, 16-19.

Aldridge, N. 2001. Little Farningham, Cranbrook – Revisited. *Archaeologia Cantiana*, V.121, 135-156.

Aldridge, N. 2009. Recent Fieldwork in Smarden, Kent. *Wealden Iron – Bulletin of the Wealden Iron Research Group*, V.2, 9-10.

Alexander, M. 2011. *Introduction to English Heritage Assets – Mills*. Swindon: English Heritage.

Allen, J. R. 2010. The Alkali-Metal Ratio in Romano-British Bloomery Slags, Severn Estuary Levels, South West Britain: Values and Implications. *Archaeology in the Severn Estuary 2009. Annual Report of the Severn Estuary Levels Research Committee.* V.20, 41-45.

Allen, J. R. and Fulford, M. G. 1999. Fort Building and Military Supply along Britain's Eastern Channel and North Sea Coasts: The Later Second and Third Centuries. *Britannia*, V.30,163-184.

Allen, J. R. and Fulford, M. G. 2004. Early Roman Mosaic Materials in Southern Britain, with Particular Reference to Silchester: A Regional Geological Perspective. *Britannia*, V.35, 9-38.

Allen, M. Post Doctorate Research Fellow, University of Reading (Presentation at Society of Antiquaries of London at the Roman Rural Settlement in the South East Conference, 30th of October 2013).

Alpin, J. 1997. The Roman Villa at Aldington. *Kent Archaeological Review*, Issue 128, 194-195.

Anderson, J. D. 1992. *Roman Military Supply in North East England.* Oxford: BAR Publishing.

Andrews, C. 2001. Romanisation: A Kentish Perspective. *Archaeologia Cantiana*, V.121, 25-43.

Andrews, P. and Buss, B. 2011. *Fairlawne Estate Roman Building Investigation: Phase 2*. Unpublished: Wessex Archaeology.

Andrews, P. Biddulph, E. Hardy, A. and Brown, R. 2011. *Settling the Ebbsfleet Valley, Sites V.1.* Oxford: Oxford Wessex Archaeology.

Applebaum, S. 1972. Roman Britain. In: Finberg, H.P.R. ed. *The agrarian history of England and Wales*. Cambridge: Cambridge University Press.

Applegate, D. 2015. 'Hoo Ware' – an investigation and comparative analysis of evidence recovered from a first-century AD Romano-British site in North Kent. *Journal of Roman Pottery Studies*. V.16, 23-53.

Arnott, S. 1862. Field Notes. Iron Works of the County of Sussex. *Sussex Archaeological Collections.* V.21. 138.

Ashbee, P. 2005. *Kent in Prehistoric Times.* Stroud: The History Press.

Atkinson, D. 1933. *Historical Essays in Honour of James Tait*. Manchester: Manchester University Press.

Austen, E. 1947. *Brede: The Story of a Sussex Parish.* Rye: Adams and son.

Bang, P. 2008. *The Roman Bazaar: A Comparative Study of Trade and Markets in Tributory Empire*. Cambridge: Cambridge University Press.

Barker, G. Webley, D. 1977. An integrated economy for Gatcombe. In: Branigan, K. ed. *Gatcombe Roman Villa*. Oxford: BAR Publishing.

Bateman, N. 2011. *Roman London's Amphitheatre*. London: Museum of London Archaeology.

de la Bédoyère, G. 1992. *Roman Towns in Britain*. London: Batsford.

de la Bédoyère, G. 1999. *The Golden Age of Britain*. Stroud: Tempus.

de la Bédoyère, G. 2000. *Pottery in Roman Britain*. Princes Risborough: Shire Publications.

de la Bédoyère, G. 2015. Face to Face with the Past: Uncovering the Real Lives of Roman Britain. *Current Archaeology*, Issue 304, V.25, 26-33.

Bell, M. 1981. Valley Sediments and Environmental Change. In: Jones, M.K. and Dimbleby, G.W. ed. *The Environment of Man: the Iron Age to the Anglo-Saxon period*. Oxford: BAR Publishing, 75-91.

Bell, T. 1999. Churches on Roman Buildings: Christian Associations and Roman Masonry in Anglo-Saxon England. *Medieval Archaeology*, V.42, 1-18.

Bennett, P., Ridler, I. and Sparey-Green, C. 2010. *The Roman Watermills and Settlement at Ickham, Kent*. Canterbury: Canterbury Archaeology Trust.

Beswick, W. 1991. Pipe Like Objects from Bloomery Sites. *Sussex Archaeological Collections*, V.129, 246-249.

Betts, I. 1987. Ceramic Building Material: Recent Work in London. *Archaeology Today*, Issue 8, 26-28.

Biddulph, E. 2013. Salt of the Earth: Roman Industry at Stanford Wharf. *Current Archaeology*, Issue 276, V.24, 16-22.

Bidwell, P. 1980. *Roman Exeter: Fortress and Town*. Exeter: Exeter City Council.

Bidwell, P. 1995. Review of Anderson 1992. *Britannia*, V.26, 395-396.

Biot, T. 2011. *A Town Unearthed* (Online). Available from: http://www.atownunearthed.co.uk/research/talks/talk-quern-stones/) [Accessed 10 January 2013].

Birley, A. R. 2005. *The Roman Government of Britain*. Oxford: Oxford University Press.

Black, E. W. 1982. The Roman Villa at Darenth. *Archaeologia Cantiana*, V.97, 159-183.

Black, E. W. 1987. *The Roman Villas of South East England*. Oxford: BAR Publishing.

Black, E. W. 2013. Roman Relief: Patterned Tiles At Dover and Elsewhere in Kent. *Kent Archaeological Review*, V.191, 40-42.

Blagg, T. 1976. Tools and Techniques of the Roman Stonemason in Britain. *Britannia*. Vol.7, 152-172.

Blagg, T. 1980. The Sculptured Stones. In: Dyson, T. ed. *The Roman Riverside Wall and Monumental Arch in London - Special Paper No 3*. London: London and Middlesex Archaeological Society, 125-193.

Blagg, T. 1990. Building Stone in Roman Britain. In: Parsons. D. ed. *Stone: Quarring and Building in England. AD 43-1525*. Chichester: Phillimore, 33-50.

Blagg, T. 2002. *Roman Architectural Ornament in Britain*. Oxford: BAR Publishing.

Blanning, E. 2008. *Towards an Interpretation of the Detached and 'Isolated' Bath Houses of Roman Kent - MA Dissertation*. Unpublished: University of Kent.

Blanning, E. 2014. *Landscape, Settlement and Materiality: Aspects of Rural Life in Kent during the Roman Period. Doctor of Philosophy (PhD) Thesis*. Unpublished: University of Kent.

Blockley, K., Blockley, M., Blockley, P., Frere, S.S. and Stowe, S. 1995. Excavations in the Marlowe Car Park and Surrounding Area - The Archaeology of Canterbury Volume 5. Canterbury: Canterbury Archaeological Trust.

Blows, J. 2011. *Strategic Stone Study – A Building Stone Atlas of Kent*. Swindon: English Heritage.

Bonifay, M. 2014. Africa: Patterns of Consumption in Coastal Regions Versus Inland Regions. The Ceramic Evidence. *Late Antique Archaeology*, V.10 (1), 529-566.

Boon, G. C. 1978. Excavations of the site of a Roman Quay at Caerleon and its Significance. In: Boon, G.C. ed. *Roman Sites*. Cardiff: Cambrian Archaeological Association Monographs and Collections.

Booth, P. 2001. The Roman Shrine at Westhawk Farm. Archaeologia Cantiana, V.121, 1-41.

Booth, P. 2012. Roman Britain in 2011 – Southern Counties. *Britannia*, V.43, 271-421.

Booth, P., Bingham, A.M. and Lawrence, S. 2008. The Roman Roadside Settlement at Westhawk Farm, Ashford, Kent. Oxford: Oxford Archaeology.

Borsos, E., Makra, L., Beczi, R., Vitanyi, B. and Szentpeter, M. 2003. Anthropogenic Air Pollution in the Ancient Times. *ACTA Climatologica Et Chorologica, Universitatis Szegediensis*, V. 36-37, 5.15.

Botting, W. J. 1973. The Romano-British Iron Working Site at Ludley Farm, Beckley. *Sussex Archaeological Collections*, V.111, 111.

Boyce, K. 2007. The Implications of Isolated Bath Houses in the Roman Cray Valley. *London Archaeologist*, V.11 Number 10, 260-264.

Bradley, P. 2013. Death Pits at Cliffs End. *British Archaeology*, Issue 131, 16-20.

Bradley, T. and Butler, J. 2008. *From Temples to Thames Street - 2000 Years of Riverside Development: Archaeological Excavations at the Salvation Army International Headquarters.* London: Pre-Construct Archaeology.

Brandon, P. 2003. *The Kent and Sussex Weald.* Bognor Regis: Phillimore and Co Ltd.

Branigan, K. 1982. Celtic Farm to Roman Villa. In: Miles, D. ed. *Romano-British Countryside.* Oxford: BAR Publishing.

Bray, I. (2010). Horrible, Speculative, Nasty and Dangerous: Assessing the Value of Roman Iron. *Britannia*, V.42, 175-185.

Breeze, D. J. and Dobson, B. 2000. *Hadrian's Wall.* London: Penguin.

Brickstock, R. and Casey, J. 2010. Roman Coins. In: Bennett, P., Ridler, I. and Sparey-Green, C. eds. *The Roman Watermills and Settlement at Ickham, Kent.* Canterbury: Canterbury Archaeology Trust, 74-84.

Briggs, A. 1994. *A Social History of Britain.* London: Weidenfeld and Nicolson.

Brigham, T., Goodburn, D., Tyers, I. and Dillon, J. 1995. A Roman Timbered Building on the Southwark Waterfront. *The Archaeological Journal*, V.152, 1-72.

Briquet, A. 1930. *Le Littoral du nord de la France et son evolution morphologique.* Paris: A.Colin.

Brodribb, G. 1969. Stamped Tiles of the Classis Britannica. *Sussex Archaeological Collections.* V.107. 102- 125.

Brodribb, G. 1970. Stamped Tiles of the Classis Britannica. *Kent Archaeological Review,* V.21, 25. Available from: http://cka.moon-demon.co.uk/KAR021/KAR021_CB.htm [Accessed 2nd December 2015].

Brodribb, G. 1979. A Survey of Tile at the Roman Bath House at Beauport Park, Battle, East Sussex. *Britannia,* V.10, 139-156.

Brodribb, G. 1980. A Further Survey of Stamped Tiles of the Classis Britannica. *Sussex Archaeological Collections.* V.118. 183 - 196.

Brodribb, G., Cleere, H., Henig, M., MacKreth, D. F. and Greep, S. J. 1988. The 'Classis Britannica' Bath-House at Beauport Park, East Sussex. *Britannia*, V.19, 217-274.

Bromwich, J. 2009. *The Roman Remains of Northern and Eastern France: A Guidebook.* London: Routledge.

Brookes, S. and Harrington, S. 2010. *The Kingdom and People of Kent, AD400 – 1066.* Stroud: The History Press.

Brooks, N. P. 1994. Rochester Bridge AD43 to 1381. In: Yates, N. and Gibson, J. H. ed. *Traffic and Politics – The Construction and Management of Rochester Bridge AD 42-1993.* Woodbridge: The Boydell Press, 1-35.

Brooks, O. J. A. 1988. The Case for Continuity in 5th Century Canterbury Re-Examined. Oxford Journal of Archaeology, V.7, 99-113.

Brown, A. G., Meadows, I., Turner, S.D. and Mattingly, D. 2001. Roman Vineyards in Britain: stratigraphic and palynological data from Wollaston in the Nene Valley, England. *Antiquity*, V.75, 745-757.

Brown, J. 2013. Enshrined by Conservation: A Romano-British Shrine in Rutland Water. *Current Archaeology*, Issue 285, V.24, 20-24.

Brown, P. 2012. *Through the Eye of a Needle.* Princeton: Princeton University Press.

Burnham, B. C., Wacher, J. 1990. *The Small Towns of Roman Britain.* London: Batsford.

Burnham, B. C. and Davies, J. L. 2010. *Roman Frontiers in Wales and the Marches.* Aberystwyth: Royal Commission on the Ancient and Historical Monuments of Wales.

Burrows, R. A. 1952. Research and Discoveries in Kent 1951 – 1952: Cuxton. *Archaeologia Cantiana*, V.65, 193.

Campbell, B. 2011. *Rivers and the Power of Ancient Rome.* Chapel Hill: University of North Carolina Press.

Canterbury Archaeological Trust. 2013. *Folkestone to 1500, A Town Unearthed.* (Online). Available from: http://www.canterburytrust.co.uk/folkestone/gallery01.htm [Accessed 23rd December 2015].

Carroll, M. 2009. Cologne. In: Gagarin, M. ed. *The Oxford Encyclopaedia of Greece and Rome.* Oxford: Oxford University Press, 251-260.

Carroll, M. 2016. Vagnari. *Current World Archaeology*, Issue 76, V.7, 30-33.

Carruthers, W. J. 2014. The Charred and Mineralised Plant Remains. In: O'Shea, L. and Weeks, J. ed. Evidence of a Distinct Focus of Romano-British Settlement at Maidstone? Excavations at Church Street 2011-12. *Archaeologia Cantiana*, V.135, 143-147.

Castle, S. A. 1978. Amphorae from Brockley Hill, 1975. *Britannia*, V.9, 383-392.

Catling, C. 2013. Chedworth Roman Villa: Life in the Cotswolds Then and Now. *Current Archaeology*, Issue 284, V.24, 43.

Catling, C. 2014. Excavating Earth: Was the Roman Fenland an Imperial Estate. *Current Archaeology*, Issue 295, V.25, 20-25.

Champion, T. 2007. Prehistoric Kent. In: Williams, J.H. ed. *The Archaeology of Kent to AD 800*. Woodbridge: The Boyden Press and Kent County Council.

Chaplin, R. E. 1962. Excavations in Rochester, Winter 1961-62. *Archaeologia Cantiana*, Vol. 77, 1-li.

Charles, T. 1847. Proceedings. *Journal of the British Archaeological Association*, V.2, 86 - 88.

Chown, E. 1947. Painted Iron Age Pottery at Seddlescombe. *Sussex Notes and Queries*, V.13, 16-19.

Chuter, G., Washington, B. and Corbett, S. 2008. An Interim Report On the Excavations at Wilbees Farm, Arlington. Available from: http://www.brightonarch.org.uk/13_9.htm [Accessed 4th November 2015].

Cleere, H. 1969. List of Sites. *Wealden Iron – Bulletin of the Wealden Iron Research Group*, V.1, 13-18.

Cleere, H. 1970. *The Romano-British Industrial Site at Bardown.* Lewes: Sussex Archaeological Society.

Cleere, H. 1974. The Roman Iron Industry in the Weald and its Connections with the Classis Britannica. *Archaeological Journal,* V.131, 17-31.

Cleere, H. 1977. *The Classis Britannica.* In: Johnston, D.E. ed. *The Saxon Shore*, York: CBA Research Report No 18, 16-19.

Cleere, H. and Crossley, D. 1995. *The Iron Industry of The Weald.* Cardiff: Merton Priory Press.

Coles, *Mr Coles Observation of Nuisances on the Medway.* 15th June 1630. *An Account of the Proceedings of the Commissioners of Sewers towards making the River Medway navigable between Maidstone and Penshurst 1627-1630*. ms. Medway Archives and Local Studies Centre. AZ1 (a late 18th century reproduction of the original, within the Medway Navigation Company's papers for the year 1800). Strood.

Connolly, P. 1981. *Greece and Rome at War.* London: Macdonald Pheobus Ltd.

Cook, N. C. 1928. A Roman Site in the Church Field at Snodland. *Archaeologia Cantiana*, V.40, 79 - 83.

Cool, H. E. M and Mason, D. J. P. 2008. *Roman Piercebridge, Excavations by D.W Harding and Peter Scott 1969 – 1981.* Durham: The Architectural and Archaeological Society of Durham and Northumberland.

Cooper, C. 2008. *Maidstone: A History*. Chichester: Phillimore.

Cornell, T. J. 1993. The End of Roman Imperial Expansion. In: Rich, J. and Shipley, G. eds. *War and Society in the Roman World.* London: Routledge, 139-170.

Cornwell, K., Cornwell, L. and Pagham, D. 2007. Castle Croft, Ninfield. *Hastings Area Archaeological Research Group*, No. 23, 3.

Cornwell, K. and Cornwell, L. 2008. A Roman Site on Kitchenham Farm, Ashburnham. *Hastings Area Archaeological Research Group*, No. 24, 10.

Cornwell, K. and Cornwell, L. 2008. A Roman Site on Kitchenham Farm, Ashburnham – 2nd Interim Report – Ceramic Building Material. *Hastings Area Archaeological Research Group*, No. 25, 1–10.

Cornwell, K. and Cornwell, L. 2010. The Stamps of the *Classis Britannica* from Kitchenham Farm, Ashburnham. *Hastings Area Archaeological Research Group*, No. 29, 16-18.

Cornwell, K. and Cornwell, L. 2013. Footlands Farm, Seddlescombe: A geophysical survey of the iron-production complex and its transport links. *Hastings Area Archaeological Research Group*, No. 33, 1 – 22.

Cotteril, J. 1993. Saxon Raiding and the Role of the Late Coastal Forts in Britain. *Britannia*, V. 24. 227-239.

Cowan, C., Seeley, F., Wardle, A., Westman, A. and Wheeler, L. 2009. *Roman Southwark, Settlement and Economy: Excavations in Southwark 1973-1991.* London: Museum of London Archaeology Monograph 42.

Cowan, R. 2015. *Roman Legionary AD 284-337*. Oxford: Osprey Publishing.

Crawford, D. J. 1976. Imperial Estates. In: Finley, M. I. ed. *Studies in Roman Property.* Cambridge: Cambridge University Press, 35-70.

Creighton, Dr. J. 30th October 2013. Director of Information and Communications Technology and Associate Professor in Archaeology, University of Reading (responding to questions. Society of Antiquaries of London, Roman Rural Settlement in the South East Conference).

Crew, P. 1998. Laxton Revisited. *Historical Metallurgy*, V.32, 49-53.

Croft, A., Munby, J. and Ridley, M. 2001. *Kent Historic Landscape Characterisation.* Maidstone. Oxford Archaeology.

Crowley, L. 2011. The role of mortuary ritual in the construction of social boundaries by privileged social groups within villa landscapes. In: Roymans, N. and Derks, T. ed. *Villa Landscapes in the Roman North.* Amsterdam: Amsterdam University Press, 195–210.

Crowley, N. and Betts, I. 1992. Three Classis Britannica Stamps from London. *Britannia*, V. 23. 218-222.

Crummy, P. 2008. The Roman Circus at Colchester. *Britannia*, V. 39, 15-31.

Cunliffe, B. 1969. Roman Kent. In: Newman, J. ed. *The Buildings of England, West Sussex and the Weald.* London: Penguin, 22-24.

Cunliffe, B. 1980. The Evolution of Romney Marsh: A Preliminary Romney Statesmen. In: Thompson, F.H.

ed. *Archaeology and Coastal Change*. London: The Society of Antiquities, 37-55.

Cunliffe, B. 1988a. *Greeks, Romans and Barbarians. Spheres of Interaction*. London: Batsford Ltd.

Cunliffe, B. 1988b. Romney Marsh in the Roman Period. In: Edisson, J. and Green, C. ed. *Romney Marsh: Evolution, Occupation, Reclamation*. Oxford: OUCA Monograph 24, 83-87. Also published as: Cunliffe, B. 1988. *Romney Marsh in Roman Period*. Rye: Romney Marsh Research Trust.

Cunliffe, B. 2009. *Iron Age Communities in Britain – An Account of England, Scotland and Wales from the 7th century BC Until the Roman Conquest*. London: Routledge.

Cunliffe, B. 2013. *Britain Begins*. Oxford: Oxford University Press.

Cunliffe, B., Reece, R., Henig, M., Chadwick Hawkes, S., Care, V. and Young, C. J. 1980. Excavations at the Roman fort at Lympne, Kent 1976-78. *Britannia*, V.11, 227-288.

Dalton, A. 1983. Field notes. *Wealden Iron – Bulletin of the Wealden Iron Research Group*, V.2, 34-35.

D'Amato, R. 2009. *Imperial Roman Naval Forces 31BC – AD500*. Oxford: Osprey Publishing.

Dance, W. 1995. Roman Coin Hoard Found at Ore in 1989. *HAARG Journal*, New series, No 1, 3-4.

Daniels, A. 2008. *An Archaeological Investigation Using Trial Pits into the Car Park and Cemetery Extension of St Margaret's Church, Barming*. Unpublished: Maidstone Area Archaeological Group.

Daniels, A. 2010. East Farleigh Roman Buildings. *Kent Archaeological Society Newsletter*. Issue 86, 12-13.

Daniels, A. 2012. East Farleigh Roman Buildings. *Kent Archaeological Review*, Issue 188, 179-183.

Daniels, A. 2015. Excavations at Oaklands, Lower Road, East Farleigh, Maidstone, 2014. *Kent Archaeological Society Newsletter*, Issue 101, 6.

Dannell, G. and Mees, A. 2015. Getting Samian Ware to Britain: Routes and Transport Possibilities. *Journal of Roman Pottery Studies*. V.16, 77-92.

Dark, K. and Dark, P. 1997. *The Landscape of Roman Britain*. Stroud: Sutton Publishing.

Darvill, T. and McWhirr, A. 1982. Roman Brick Production and the Environment. In: Miles, D. ed. *Romano-British Countryside*. Oxford: BAR Publishing, 137-150.

Davies, J. 1982. A Hoard of 'Radiate' Coins from Allington. *Archaeologia Cantiana*, V.98, 137-144.

Davies, M. 2009. The Evidence of Settlement at Plaxtol in the Late Iron Age and Romano-British Periods. *Archaeologia Cantiana*, V.129, 257-278.

Davies, N. 2011. *Vanished Kingdoms*. London: Penguin.

Dawkes, G. 2009. *A Post Excavation Assessment Report and Updated Project Design on Archaeological Excavations at High Street, Snodland, Kent*. Portslade: ASE.

Defoe, D. 1724. *Tour Thro' the Whole Island of Great Britain – Vol 1, Letter II*. London: Penguin.

Detsicas, A. 1967. Excavations at Eccles Roman Villa, 1966. Fifth Interim Report. *Archaeologia Cantiana*, V.82, 170-178.

Detsicas, A. 1977. First Century Pottery Manufacture at Eccles, Kent. In: Dore, J. and Greene, K. eds. *Roman Studies in Britain and Beyond*. Oxford: BAR Publishing, 19-36.

Devoy, R. J. N. 1990. Controls on Coastal and Sea-Level Changes and the Application of Archaeological Historical Records to Understanding Recent Patterns of Sea-Level Movement. In: McGrail, S. ed. *Maritime Celts, Frisians and Saxons*. York: CBA Research Group, 17-26.

Driessen, M. and Besselsen, E. 2014. *Voorburg-Arentsburg. Een Romeinse Havenstad tussen Rijn en Mass*. Amsterdam, University of Amsterdam.

Dumayne-Peaty, L. 1998. Forest Clearance in Northern Britain during Romano-British Times: Re-addressing the Palynological Evidence. *Britannia*, V.29, 315-322.

Durham, A. and Goormachtigh, M. 2015. Greenwich and the Early Emporia of Kent. *Archaeologia Cantiana*, V.136, 163-176.

Dworakowska, A. 1983. *Quarries in Roman Provinces*. Wroclaw: Zaklad Narodowy im. Ossolinskich.

Elliott, P. 2014. *Legions in Crisis*. Stroud: Fonthill Media ltd.

Elliott, S. 2011. *The Medway Formula – MA Dissertation*. Unpublished: University College London.

Elliott, S. 2013. A Roman villa At Teston. *British Archaeology*, Issue 133, 40-46.

Elliott, S. 2014a. The Medway Formula - A Search for Evidence That the Roman Authorities Improved the River's Navigability to Facilitate Their Extensive Ragstone Quarrying Industry. *Archaeologia Cantiana*, V.135, 251-260.

Elliott, S. 2014b. Britain's First Industrial Revolution. *History Today*, Issue 5, V.64, 49-55.

Elliott, S. 2014c. The Mystery of the Medway Stones. *Current Archaeology*, Issue 298, V.25, 11.

Elliott, S. 2016. *Sea Eagles of Empire: The Classis Britannica and the Battles for Britain*. Stroud: The History Press.

Elliott, S. 2017. *'Change and continuity in the exploitation of natural resources (such as stone, iron, clay and wood) in the principal areas of industrial activity in Kent (namely the Weald, Folkestone region and upper Medway Valley) during the Roman occupation'.* Unpublished. University of Kent.

Ellis Jones, J. 2012. *The Maritime Landscape of Roman Britain.* Oxford: BAR Publishing.

English, J. 2012. Roman Curse Scroll from East Farleigh, Kent. CBA South East Newsletter, Issue 38, 3-4.

English Heritage. 2011. *Strategic Stone Study – A building Stone Atlas of Kent.* Swindon: English Heritage.

Erdkamp, P. 2005. *The Grain Market in the Roman Empire: A Social, Political, and Economic Study.* Cambridge: Cambridge University Press.

Erdkamp, P. 2013. The Food Supply of the Capital. In: Erdkamp, P. eds. *The Cambridge Companion to Ancient Rome.* Cambridge: Cambridge University Press. 262-278.

Esmonde Cleary, A. S. 1989. *The Ending of Roman Britain.* London: Batsford.

Esmonde Cleary, A. S. 2013. *The Roman West AD 200 – 500.* Cambridge: Cambridge University Press.

Evans, C. and Newman, R. 2014. *North West Cambridge: University of Cambridge. Archaeological Evaluation Fieldwork.* Cambridge: Cambridge Archaeological Unit.

Evans, J. H. 1953. Archaeological Horizons in the North Kent Marshes. *Archaeologia Cantiana,* V.66, 103-146.

Evans, J. 2013. Balancing the Scales: Romano-British Pottery in Early Late Antiquity. *Late Antique Archaeology,* V.10 (1), 425-450.

Everitt, A. 1986. *Continuity and Colonisation.* Leicester: Leicester University Press.

Farebrother, G. 1978. Bloomery at Rushlake Green. *Wealden Iron – Bulletin of the Wealden Iron Research Group,* V1, 5.

Faulkner, N. 2001. *The Decline and Fall of Roman Britain.* Stroud: Tempus.

Fernando-Lozano, J., Gutierrez-Alonso, G. and Fernandez-Moran, M. A. 2015. Using Airborne LIDAR Sensing Technology and Aerial Orthoimages to Unravel Roman Water Supply Systems and Gold Works in NW Spain. *Journal of Archaeological Science,* V. 53, 356-373.

Fields, N. 2006. *Rome's Saxon Shore.* Oxford: Osprey Publishing.

Fincham, G. 2002. *Landscapes of Imperialism: Roman and native interaction in the East Anglian Fenland.* Oxford: BAR Publishing.

Finkel, I. 2014. *The Ark Before Noah.* London: Hodder and Stoughton.

Finley, M. 1999. *The Ancient Economy.* Berkely: University of California Press.

Fisher, C. E. 1939. Reports. *Archaeologia Cantiana,* V.51, 204.

Fleming, R. 2010. *Britain After Rome.* London: Penguin.

Foreman, S. 1999. *Thurnham Roman Villa, Thurnham, Kent.* Unpublished: Oxford Archaeological Unit.

Frere, S. 1987. *Britannia: A History of Roman Britain* (3rd edn). London: Routledge.

Fuhrmann, C. J. 2011. *Policing the Roman Empire: Soldiers, Administration and Public Order.* Oxford: Oxford University Press.

Fulford, Prof. M. 30th October 2013. Presentation at Society of Antiquaries of London at the Roman Rural Settlement in the South East Conference.

Gardiner, M. 1996. *An Archaeological Evaluation of Land adjacent to Pested Bars Road, Boughton Monchelsea, Maidstone, Kent.* Unpublished: South Eastern Archaeological Service.

Gardner, A. 2007. *An Archaeology of Identity.* California: Left Coast Press.

Gardner, A. 2013. Thinking about Roman Imperialism: Postcolonialism, Globalisation and Beyond? *Britannia,* V.44, 1-25.

Gibson-Hill, J. 1972. The Rescue Excavation at Broadfields. *Wealden Iron – Bulletin of the Wealden Iron Research Group,* V1, 25-26.

Goacher, D. 2009. The Population Size of Romano-British Kent: An Initial Estimate. *Archaeologia Cantiana,* V.129, 391-398.

Goacher, D. 2012. Kentish Ragstone from the Maidstone Area. *Kent Archaeological Society Newsletter,* Issue 94, 12-13.

Godfrey, J. 1962. *The Church in Anglo-Saxon England.* Cambridge: Cambridge University Press.

Goldsworthy, A. 2003. *The Complete Roman Army.* London: Thames and Hudson.

Goldsworthy, A. 2014. *Augustus: From Revolutionary to Emperor.* London: Weidenfeld and Nicolson.

Goodburn, D. 1991. A Roman Timber Framed Building Tradition. *The Archaeological Journal,* V.148, 182-204.

Grainge, G. 2005. *The Roman Invasions of Britain.* Stroud, Tempus.

Gray, H. 1915. *English Field Systems.* Harvard University Press: Harvard.

Green, C. 2010. Forest Ports of the Seven. *Archaeology in the Severn Estuary 2009.* V.20, 47-63.

Green, C. 2013. Quernstone Production Around Copt Point: Folkestone's First Industry. In: Coulson, I. ed. *Folkestone to 1500, A Town Unearthed.* Canterbury: Canterbury Archaeological Trust. 50-51.

Green, C. 2014. Querns and Millstones in Late Iron Age and Roman London and South-East England. Unpublished: *in prep*.

Green, C. and Peacock, D. 2011, *The Puddingstone Industries in France and Britain.* (Online). Available from: http://www.sal.org.uk/fundraising/research/puddingstone/ [Accessed 12th November 2012].

Green, C. and Peacock, D. 2012. Worms Heath Puddingstone Querns: Crawford Vindicated. *Surrey Archaeological Society Bulletin*, Issue 436, 1-3.

Greene, K. 1986. *The Archaeology of the Roman Economy.* Berkeley and Los Angeles: University of Chicago Press.

Griffin, E. 2010. *A Short History of the British Industrial Revolution.* Basingstoke: Palgrave Macmillan.

Grover, J. W. 1873. Notes on the Foundations of a Roman Villa at Teston, Kent. *The Journal of the British Archaeological Association*, V.29, 45-47.

Guest, P. and Young, T. 2009. Mapping Isca: Geophysical Investigation of School Field and Priory Field, Caerleon. *Archaeologia Cambrensis*, Issue 158, 97-111.

Halkon, P. 2011. Iron Landscape and Power in Iron Age East Yorkshire. *The Archaeological Journal*. V.168. 134-165.

Hall, J. and Merrifield, R. 1986. *Roman London*. London: HMSO Publications for the Museum of London.

Halsall, G. 2013. *Worlds of Arthur.* Oxford: Oxford University Press.

Hann, Dr A. 2008. The Lower Medway Valley. Local History and Community Involvement Presentation for Victoria County History's 'England's Past for Everyone', sponsored by English Heritage. (Online). Available from: http://www.google.co.uk/url?sa=t&rct=j&q=&esrc=s&source=web&cd=1&ved=0CCIQFjAA&url=http%3A%2F%2Fwww.english-heritage.org.uk%2Fcontent%2Fimported-docs%2Fpower-point-presentations%2Flower-medway-valley-local-history&ei=qsdlVaj5Lcf8UOnCgcAK&usg=AFQjCNHFch8YziEQayQR4lzVzMT_e9Ux7g&bvm=bv.93990622,d.d24 [Accessed 18th July 2013].

Hanson, W.S. 1996. Forest Clearance and the Roman Army. *Britannia*, V.27, 354-358.

Harrington, S. and Welch, M. 2014. *The Early Anglo-Saxon Kingdoms of Southern Britain.* Oxford: Oxbow Books.

Harris, O. 2008. John Leland and the 'Briton Brykes'. *The Antiquiries Journal*, V.87, 346-356.

Harrison, A.C. and Flight, C. 1968. The Roman and Medieval Defences of Rochester in Light of Recent Excavations. *Archaeologia Cantiana*, V.83, 55-104.

Haselgrove, C. 1987. Culture Process on the Periphery: Belgic Gaul and Rome during the late Republic and early Empire. In: Rowlands, M., Larsen, M. and Kristiansen, K. eds. *Centre and Periphery in the Ancient World.* Cambridge: Cambridge University Press. 104-124.

Haselgrove, C. 1997. Iron Age Brooch Deposition and Chronology. In: Gwilt, A. and Haselgrove, C. ed. *Reconstructing Iron Age Society*. Oxford: Oxbow Monograph 71, 51-72.

Haselgrove, C. 2005. A New Approach to Analysing the Circulation of Iron Age Coinage. *The Numismatic Chronicle*, V.165, 129-174.

Haslam, J. 2010. *Early Medieval Towns in Britain*. Princes Risborough: Shire Archaeology.

Hasted, E. 1797. *The History of Topographical Survey of the County of Kent - East Barming, Teston and West Barming*. Ashford: 2008 reprint by Old Towns Books and Maps.

Hasted, E. 1798a. *The History of Topographical Survey of the County of Kent - The Town and Parish of Rye*. Ashford: 2008 reprint by Old Towns Books and Maps.

Hasted, E. 1798b. *The History of Topographical Survey of the County of Kent - Addington, Birling, Offham, Ryarsh, Trottiscliffe.* Ashford: 2008 reprint by Old Towns Books and Maps.

Hasted, E. 1800. *The History of Topographical Survey of the County of Kent - East and West Farleigh*. Ashford: 2008 reprint by Old Towns Books and Maps.

Hastings, P. 2000. *Upon the Quarry Hills – A History of Boughton Monchelsea.* Maidstone: Boughton Monchelsea Parish Council.

Haverfield, F. J. 1916. Field Notes. *Sussex Archaeological Collections.* V.58. 159.

Haverfield, F. J. and Wheeler, R. E. M. 1932. Introduction. In: Page, W. ed. *The Victoria History of the Counties of England – Kent.* London: St. Catherine Press, 144-177.

Hayward, K. M. J. 2009. *Roman Quarrying and Stone Supply on the Periphery - Southern England.* Oxford: BAR Publishing.

Heather, P. 2013. *The Restoration of Rome.* London: Macmillan.

Helm, R. and Carruthers, W. 2011. Early Roman Evidence for Intensive Cultivation and Malting of Spelt Wheat at Nonington. *Archaeologia Cantiana*, V.131, 353-372.

Henig, M. 1995. *The Art of Roman Britain*. London: Routledge.

Henson, D. 2006. *The Origins of the Anglo-Saxons.* Ely: Anglo-Saxon Books.

Hewitt, E. M. 1932. Industries. In: Page, W. ed. *The Victoria History of the Counties of England – Kent.* London: St. Catherine Press, 371-431.

Hill, J. D. 1997. The End of One Kind of Body and the Beginning of Another Kind of Body. In: Gwilt, A. and

Haselgrove, C. ed. *Reconstructing Iron Age Society.* Oxford: Oxbow Monograph 71, 96-107.

Hill, P. R. and David, J. C. E. 1995. *Practical Stone Masonry.* Shaftsbury: Donhead Publishing.

Hingley, R. 1982. Roman Britain: The structure of Roman imperialism and the consequences of imperialism on the development of a peripheral province. In: Miles, D. ed. *The Romano-British Countryside: Studies in Rural Settlement and Economy*. Oxford: BAR Publishing, 17-52.

Hingley, R. 1989. *Rural Settlement in Roman Britain.* London: Seaby.

Hingley, R. 2005. *Globalizing Roman Culture – Unity, Diversity and Empire.* London: Routledge.

Hingley, R. 2007. The Roman Landscape of Britain: From Hoskins to Today. In: Fleming, A. and Hingley, R. eds. *Prehistoric and Roman Landscapes.* Oxford: Windgather Press, 101-112.

Higgins, C. 2013. Under Another Sky: Journeys in Roman Britain. London: Jonathan Cape.

Hirt, A.M. 2010. *Imperial mines and Quarries in the Roman World.* Oxford: Oxford University Press.

Hodges, R. 2014. Roman Goldmines in Transylvania. *Current World Archaeology*, Issue 68, 54-57.

Hodgkinson, J. 1987. Footlands Iron Working Site, Seddlescombe. *Wealden Iron – Bulletin of the Wealden Iron Research Group*, V.2, 22-32.

Hodgkinson, J. 1988. Two Roman Shoe Fragments from Seddlescombe. *Sussex Notes and Queries*, V.126, 231-233.

Hodgkinson, J. 1988. Field Notes. *Wealden Iron – Bulletin of the Wealden Iron Research Group*, V.2, 2-11.

Hodgkinson, J. 1991. Field Notes. *Wealden Iron – Bulletin of the Wealden Iron Research Group*, V.2, 2-7.

Hodgkinson, J. 1992. Field Notes. *Wealden Iron – Bulletin of the Wealden Iron Research Group*, V.2, 8-9.

Hodgkinson, J. 1993. Field Notes. *Wealden Iron – Bulletin of the Wealden Iron Research Group*, V.2, 2-3.

Hodgkinson, J. 1994. Field Notes. *Wealden Iron – Bulletin of the Wealden Iron Research Group*, V.2, 2-3.

Hodgkinson, J. 1997. Field Notes. *Wealden Iron – Bulletin of the Wealden Iron Research Group*, V.2, 2-9.

Hodgkinson, J. 1999. Romano-British Iron Production in the Sussex and Kent Weald: a review of current data. *The Journal of the Historical Metallurgy Society*, V.33, 68-72.

Hodgkinson, J. 2001. Field Notes. *Wealden Iron Research Group Bulletin,* V.2, 3-5.

Hodgkinson, J. 2004. Field Notes. *Wealden Iron Research Group Bulletin,* V.2, 2-6.

Hodgkinson, J. 2008. *Waste as a Potential Indicator of Regional Iron Production and Organisation: An example from Southeast Roman Britain.* Unpublished, Wealden Iron Research Group.

Hodgkinson, J. 2009. *The Wealden Iron Industry.* Stroud: The History Press.

Hodgkinson, J. 2012. *Winter Meeting 2012. WIRG Newsletter 55.* (Online). Available from: http://www.wealdeniron.org.uk/Newsletters/Mar%202012%20col.pdf [Accessed 1st December 2015].

Hodgkinson, J. 16th November 2013. The Landscape of Iron Production in the Roman Weald - Presentation at CBA South East Conference on Landscapes of South Eastern Britain During the Roman Period.

Hodgson, N. 2007. Arbeia Roman Fort – South Shields. *Current Archaeology*, Issue 133, V.18, 24-42.

Holbrook, N. 2006. The Roman Period. In: Holbrook, N. and Jurica, J. eds. *Twenty Five Years of Archaeology in Gloucestershire. A Review of New Discoveries and New Thinking in Glouctershire, South Glouctershire and Bristol 1979 – 2004. Cotswold Archaeology and Bristol Archaeology, Report 3*. Cirencester, Cotswold Archaeology, 97-131.

Holman, D. 2000. Iron Age Coinage in Kent. *Archaeologia Cantiana,* V.120, 205-233.

Holman, D. 2005. Iron Age Coinage and Settlement in East Kent. *Britannia*, V.36, 1-54.

Hopkins, K. 1985. *Death and Renewal.* Cambridge: Cambridge University Press.

Houliston, M. 1999. Excavations at the Mount Villa, Maidstone. *Archaeologia Cantiana*, V.119, 71-172.

Howell, I. 2014. Continuity and Change in the late Iron Age-Roman Transition Within the Environs of Quarry Wood Oppidum: Excavations at Furfield Quarry, Boughton Monchelsea. *Archaeologia Cantiana*, V.134, 37-66.

Howell, I., Blackmore, L., Phillpotts, C. and Thorp, A. 2013. *Roman and Medieval Development South of Cheapside. Excavations at Bow Bells House, City of London, 2005-6*, London: Museum of London Archaeology, Archaeology Studies Series 26.

Hughes, I. 2010. *Stilicho: The Vandal Who Saved Rome.* Barnsley: Pen and Sword Books.

Hughes, J. 1851. On the pneumatic method adopted in constructing the foundations of the new bridge across the Medway - Minutes of the Proceedings of the Institution of Civil Engineers. In: Yates, N. and Gibson, J.M. ed. *Traffic and Politics – The Construction and Management of Rochester Bridge AD43 -1993.* Woodbridge: The Boydell Press, 9.

Hutchinson, J.N., Poole, C., Lambert, N. and Bromhead, E.N. 1985. Combined Archaeological and Geotechnical

Investigations of the Roman Fort at Lympne, Kent. *Britannia*, V.16, 88-89.

Ingle, C. J. 1989. *Characterisation and Distribution of Beehive Querns in Eastern England,* Unpublished PhD thesis: University of Southampton.

Ingle, C. J. 1993. The Quernstones from Hunsburry Hillfort, Northamptonshire. *Northamptonshire Archaeology*, V.25. 21-35.

Ingleton, R. 2012. *Fortress Kent, the Guardian of England.* Barnsley: Pen and Sword Books.

Jackson, R. 2012. *Ariconium, Herefordshire: an Iron Age settlement and Romano-British small town.* Oxford: Oxbow Books.

Jackson, R.P.J., and Potter, T.W. 1996. *Excavations at Stonea, Cambridgeshire, 1980-1985.* London: British Museum Press.

James, S. 1999. *The community of the soldiers: a major identity and centre of power in Roman Britain.* In: Baker, P., Jundi, S., and Witcher, R. eds. *TRAC 98: Proceedings of the Eighth Annual Theoretical Roman Archaeology Conference, Leicester 1998.* Oxford: Oxbow, 14-25.

James, S. 2011. *Rome and the Sword.* London: Thames and Hudson.

Jamieson, D. 2008. *PB Site, Horsewash Lane, Rochester.* Unpublished: Archaeology South East.

Jenkins, R. C. 1876. On a Roman Hypocaust Discovered at Folkestone, A.D. 1875. *Archaeologia Cantiana*, V.10, 173-177.

Jessup, R. F. 1932. Romano-British Remains – Industries. In: Page, W. ed. *The Victoria History of the Counties of England – Kent.* London: St. Catherine Press, 127-133.

Jessup, R. F. 1956. The 'Temple of Mithras' at Burham. *Archaeologia Cantiana*, V.70, 168-172.

Jobson, J. J. 1921. The Roman Bidge at Rochester. *Archaeologia Cantiana*, V.35, 138-144.

Johnston, D. E. 2004. *Roman Villas.* Princes Risborough: Shire Archaeology.

Jones, A. H. M. 1953. Inflation Under the Roman Empire. *Economic History Review,* Issue 5, 293-318.

Jones, B. and Mattingly, D. 1990. *An Atlas of Roman Britain.* Oxford: Oxbow Books.

Jones, G. D. B. 1980. The Roman Mines at Rio Tinto. *The Journal of Roman Studies*, V.70, 146-164.

Jones, G. 1981. Field Reports 1980-81. *Recologea Papers*, V.7, 68-72.

Jones, H. A. 1992. A Survey of Roman Fragments in Churches in S.E. Kent. *Kent Archaeological Review*, Issue.110, 230.

Jones, M. 1982. Crop Production in Roman Britain. In: Miles, D. ed. *Romano-British Countryside.* Oxford: BAR Publishing, 97-107.

Jones, M. 1986. Towards a Model of the Villa Estate. In: Miles, D. ed. *Archaeology at Barton Court Farm, Abingdon, Oxon. Council for British Archaeology Research Report*, V.50, 38-42.

Kaminski, J. 1994. The Environmental Implications of Romano-British Iron Production in the Weald – *PhD Thesis*. Unpublished: University of Reading.

Kaye, S. 2015a. The Roman Invasion of Britain AD, 43. Riverine, Wading and Tidal Studies Place Limits on the Location of the Two Day River Battle and Beachead. (Online). Available from: https://www.academia.edu/13579532/The_Roman_Invasion_of_Britain_AD_43_riverine_wading_and_tidal_studies_place_limits_on_the_possible_locations_of_the_two-day_river_battle_and_beachhead [Accessed 1st February 2016].

Kaye, S. 2015b. The Roman Invasion of Britain AD, 43. Riverine, Wading and Tidal Studies Place Limits on the Location of the Two Day River Battle and Beachead. *Archaeologia Cantiana*, V.136, 227-240.

Keevill, G. 2015. Rewriting Rochester: A Lost Cathedral Revealed. *Current Archaeology*, Issue 307, V.26, 18-24.

Keith, A. and Ward, G. 1934. The Snodland Skeleton. *Archaeologia Cantiana*, V.46, 202-203.

Keller, P. T. 1988. The Evidence for Ancient Quern Production at Folkeston. *Kent Archaeological Review*, Issue 99, 206-209.

Keller, P. T. 1989. Quern Production at Folkestone, South East Kent. An Interim Note. *Britannia*, V.20, 193-200.

Kelly, D. B. 1984. Researches and Discoveries in Kent. *Archaeologia Cantiana*, V.101, 349-380.

Kemble, J. M. 1856. Proceedings. *Archaeological Journal*, V.13, 403-404.

Kent County Council. 2004. Kent Historic Towns survey (Online). Available from: http://archaeologydataservice.ac.uk/archiveDS/archiveDownload?t=arch-459-1/dissemination/pdf/Hythe.pdf [Accessed 12th December 2015].

Killock, D. 2014. *An Assessment of an Archaeological Excavation at Tabard Square, 34-70 Long Lane and 31-47 Tabard Street, London SE1, London Borough of Southwark.* Unpublished: PCA Ltd.

Kissick, E. 1990. Roman Building on Bluebell Hill: Where Is It? *Kent Archaeological Review*, Issue 102, 10-14.

Kolb, A. 2001. The Cursus Publicus. In: Adams, C. and Laurence, R. eds. *Travel and Geography in the Roman Empire.* London: Routledge, 95-106.

Krier, J. and Henrich, P. 2011. Monumental funerary structures of the 1st to the 3rd Centuries associated with Roman villas in the area of the Treveri. In: Toymans, N. and Derks, T. eds. *Villa Landscapes in the Roman North.* Amsterdam: Amsterdam University Press, 211-234.

Lambarde, W. 1576 (1826 edition). *A Perambulation of Kent.* London: Adams and Dart.

Laurence, R. 2012. *Roman Archaeology for Historians.* Abingdon: Routledge.

Laurence, R. 2013. Road To Success – How Money, Mules and Milestones Led to Roman Globalisation. *Current World Archaeology*, Issue 59, 42-46.

Lavan, L. 2014. Local Economies in Late Antiquity? Some Thoughts. *Late Antique Archaeology*, V.10 (1), 1-11.

Lawson, T. and Killingray, D. 2004. *An Historical Atlas of Kent.* Chichester: Phillimore and Co Ltd for the Kent Archaeological Society.

Le Bohec, Y. 2000. *The Imperial Roman Army.* London: Routledge.

Lebon, C. 1957. Researches and Discoveries in Kent. *Archaeologia Cantiana*, V.71, 224.

Lebon, C. 1958. Field Work. *Archaeologia Cantiana*, V.72, xivii.

Lebon, C. 1961. Field Work. *Archaeologia Cantiana*, V.76, xiviii.

Legg, R. 1986. *Exploring the Heartland of Purbeck.* Sherborne: Dorset Publishing Company.

Lemak, J. 2006. Review of 'The Grain market in the Roman Empire: A Social, Political, and Economic Study'. (Online). Available from: http://bmcr.brynmawr.edu/2006/2006-12-09.html [Accessed 10th November 2016].

Lendering, J. 2002, Revision 4 October 2014). *Laurum.* (Online). Available from: http://www.livius.org/articles/place/laurum-woerden/ [Accessed 11th January 2015].

Ling, R. 1985. The Mechanics of the Building Trade. In: Grew, F. and Hobley, B. ed. *Roman Urban Topography in Britain and the Western Empire.* York: CBA Research Report No 59, 14-27.

Locker, A. 2007. *In piscibus diversis: the Bone Evidence for Fish Consumption in Roman Britain. Britannia*, V.38, 141-180.

Locker, A. 2014. The Fish Remains. In: O'Shea, L. and Weeks, J. ed. Evidence of a Distinct Focus of Romano-British Settlement at Maidstone? Excavations at Church Street 2011-12. *Archaeologia Cantiana*, V.135, 141-142.

Loe, L., Boyle, A.,Webb, H. and Score, D. 2014. *Given to the Ground: A Viking Age Mass Grave on Ridgeway Hill, Weymouth.* Oxford: Oxford Archaeology.

Lott, G. and Cameron, D. 2005. *The Building Stones of South East England; Minerology and Provenance.* Nottingham: British Geological Survey.

Lower, M. A. 1849. Iron Works of the County of Sussex. *Sussex Archaeological Collections.* V.2. 171-174.

Lucas, B. H. 1950. The Bynes Farm Romano-British Bloomer. *Sussex Notes and Queries*, V.13, 16-19.

Lyne, M. 1994. *Late Roman Handmade Wares in South East Britain – PhD Thesis.* Unpublished: University of Reading.

Lyne, M. September 1996. *Roman Ship Fittings From Richborough: Speech at ROMEC X, Montpelier.* (Online). Available from: http://www.classis-britannica.co.uk/menu/shipping/shipping.htm [Accessed 10th October 2011].

Lyne, M. 2000. *Roman Wight. Isle of Wight History Centre.* (Online). Available from: http://www.iwhistory.org.uk/resourceassessment/iow%20Roman.pdf [Accessed 12th December 2012].

McGrail, S. 2014. *Early Ships and Seafaring: European Water Transport.* Barnsley: Pen and Sword Archaeology.

McOmish, D. 2011. *Introduction to History Assets – Oppida.* Swindon: English Heritage, 2011), 359-377.

McRae, S. G. and Burnham, C. P. 1973. *The Rural Landscape of Kent.* London: Wye College.

McWhirr, A. and Viner, D. 1978. The Production and Distribution of Tiles in Roman Britain with Particular Reference to the Cirencester Region. *Britannia*, V.9, 359-377.

Magness, J. 2011. A reconsideration of Josephus' testimony About Masada. In: Popovi, m. ed. *The Jewish Revolt Against Rome: Interdisciplinary Perspectives.* Leiden: Brill, 343-361.

Malim, T. 2005. *Stonea and the Roman Fenlands.* Stroud: Tempus.

Manley, J. 2002. *AD43: The Roman Invasion of Britain.* Stroud: Tempus.

Margary, I. D. 1933. Objects Found at the Ironworks, Ridge Hill, East Grinstead. *Sussex Notes and Queries.* V.6. 177-178.

Margary, I. D. 1955. An Old Map of Cansiron in East Grinstead and Hartfield 1724. *Sussex Notes and Queries*, V.13, 100-102.

Margary, I. D. 1967. *Roman Roads in Britain.* London. John Baker Publishers Ltd.

Marsden, P. 1994. Ships of the Port of London, 1st to 11th Centuries. Swindon: English Heritage.

Mason, D. J. P. 2001. *Roman Chester: City of the Eagles.* Stroud: Tempus Publishing Ltd.

Mason, D. J. P. 2003. *Roman Britain and the Roman Navy.* Stroud: The History Press.

Mason, S. 1998. Land Adjacent to Pested Bars Road, Boughton Monchelsea, Maidstone, Kent – An Archaeological Evaluation. Unpublished: Museum of London Archaeology Service.

Matthews, J. D. 1989. *Silvicultural Systems.* Wotton-under-Edge: Clarendon Press.

Mattingly, D. 2006. *An Imperial Possession, Britain in the Roman Empire.* London: Penguin.

Mattingly, D. 2009. The Imperial Economy. In: Potter, D.S. ed. *Companion to the Roman Empire.* Hoboken: Wiley-Blackwell.

Mattingly, D. 2011. *Imperialism, Power and Identity – Experiencing the Roman Empire.* Princeton: Princeton University Press.

Mayfield, A. 2015. *Randall Manor Season 10, Shorn Woods Country Park, Kent.* (Online). Available from: http://www.dayofarchaeology.com/randall-manor-season-10/ [Accessed 18th January 2016].

Merrifield, R.1965. *The Roman City of London.* London: Ernest Benn Ltd.

Miles, A, 1972. Romano-British Buildings at The Mount, Maidstone. *Archaeolgia Cantiana,* V.87, 217-219.

Millett, M. 1990a. *The Romanization of Britain.* Cambridge: Cambridge University Press.

Millett, M. 1990b. Introduction: London as a Capital. In: Watson, B. ed. *Roman London: Recent Archaeological Work.* Portsmouth: Journal of Roman Archaeology Supplementary Series Number 24, 7-12.

Millett, M. 1995. *Roman Britain.* London: Batsford.

Millett, M. 2007. Roman Kent. In: Williams, H. ed. *The Archaeology of Kent to 800AD.* Woodbridge: The Boyden Press and Kent County Council, 135-186.

Millett, M. 2015. Broader Perspectives on Past Lives. In: Halkon, P., Millett, M., and Woodhouse, H. eds. *Hayton, East Yorkshire: archaeological studies of the Iron Age and Roman landscapes. Volumes 1 and 2. (Yorkshire Archaeological Report 7).* Leeds: Yorkshire Archaeological Society, 542-558.

Mills, P. J. E., 2013. The Supply and Distribution of Ceramic Building Material In Roman Britain. In: Levan, L. and Mulryan, M. eds. *Field Methods and Techniques in Late Antique Archaeology 10.* Leiden: Brill, 451-470.

Milne, G. 2000. A Roman Provincial Fleet; The Classis Britannica reconsidered. In: Oliver G., Brock, R., Cornell, T. and Hodgkinson, S. eds. *The Sea in Antiquity.* Oxford: BAR Publishing, 127-131.

Monaghan, J. 1987. *Upchurch nd Thameside Roman Pottery: Ceramic Typology for Northern Kent, First to Third Centuries AD.* Oxford: BAR Publishing.

Money, J. H. 1979. Excavations in the Iron Age Hill Fort and Romano-British Ironworking Settlement at Garden Hill, Hartfield, East Sussex (1968 -1978). *Wealden Iron – Bulletin of the Wealden Iron Research Group,* V1, 16-26.

Money, J. H., Fulford, M.G., and Eade, C. 1977. The Iron-Age Hill-Fort and Romano-British Iron-Working Settlement at Garden Hill, Sussex: Interim Report on Excavations, 1968-76. *Britannia,* V.8, 339-350.

Moody, G. 2008. *The Isle of Thanet. From Prehistory to the Norman Conquest.* Stroud: Tempus.

Moore, F. G. 1950. Three Canal Projects, Roman and Byzantine. *American Journal of Archaeology,* V.54, 97-111.

Moorhead, S. 2014. *A History of Roman Coinage in Britain.* Witham: Greenlight Publishing.

Moorhead, S. and Stuttard, D. 2012. *The Romans Who Shaped Britain.* London: Thames and Hudson.

Moorhead, S., Anderson, I. and Walton, P. 2012. *The Roman Coins from the Excavation at Whitefriars, Canterbury.* Unpublished: British Museum.

Morris, F.M. 2010. *North Sea and Channel Connectivity during the Late Iron Age and Roman Period (175/150BC – AD 409).* Oxford: BAR Publishing.

Morris, P. 1979. *Agricultural Buildings in Roman Britain.* Oxford: BAR Publishing.

Mould, Q. 2010. Military Equipment. In: Bennett, P., Ridler, I. and Sparey-Green, C. eds. *The Roman Watermills and Settlement at Ickham, Kent.* Canterbury: Canterbury Archaeology Trust, 141-148.

Mynott, E. 1978. A Roman Stone Coffin at Keston. (Online). Available from: http://cka.moon-demon.co.uk/KAR054/KAR054_Keston.htm [Accessed 15th July 2013].

Neilson, R. 2000. *Land Adjacent to Pested Bars Road, Bougton Monchelsea, Maidstone, Kent – A Post-Excavation Assessment and updated Project Design.* Unpublished: Museum of London Archaeology Service.

Newman, J. 1969. *West Kent and the Weald.* New Haven: Yale University Press.

Nightingale, M. D. 1952. A Roman Land Settlement Near Rochester. *Archaeologia Cantiana,* V.65, 150-159.

Oldham, P. 1977. Researches and Discoveries in Kent. *Archaeologia Cantiana,* V.93, 224.

Papworth, M. 2015. Chedworth Roman Villa. *Current Archaeology,* Issue 305, V.26, 12-18.

Parfitt, K. 2004. The Roman Building at Harp Wood, Saltwood, Near Hythe. *Kent Archaeological Review,* Issue 158, 169-171.

Parfitt, K. 2011. The Rocky Road to the Iron Age: Excavations at Folkestone Roman Villa, 2011. *Kent Archaeological Society Newsletter,* Issue 92, 2-4.

Parfitt, K. 2013. Folkestone During the Roman Period. In: Coulson, I. ed. *Folkestone to 1500, A Town Unearthed*. Canterbury: Canterbury Archaeological Trust, 31-54.

Parfitt, K. and Philp, B. 1981. Building Materials. In: Philp, B. ed. *The Excavations of the Roman Forts of the Classis Britannica at Dover, 1970-77,* Dover: Kent Archaeological Rescue Unit, 175-176.

Parker, P. 2009. *The Empire Stops Here.* London: Jonathan Cape.

Payne, G. 1880. Roman Remains Found at Chatham and at Barming. *Archaeologia Cantiana*, V.13, 168-172.

Payne, G. 1893. *Collectanea Cantiana*. London: Mitchell and Hughes.

Payne, G. 1895a. Roman Rochester. *Archaeologia Cantiana*, V.21, 1-17.

Payne, G. 1895b. Roman Building in Burham. *Proceedings of the Society of Antiquiries of London,* V.15, 184 -185.

Payne, G. 1897. Researches and Discoveries in Kent. *Archaeologia Cantiana*, V.22, lviii - lxxii.

Payne, G. 1898. Roman Discoveries. *Archaeologia Cantiana*, V.23, 1-23.

Payne, G. 1902. Researches and Discoveries in Kent. *Archaeologia Cantiana*, V.25, lxiv - lxxii.

Paynter, S. 2007. Romano-British Workshops for Iron Smelting at Westhawk Farm, Kent. *Historical Metallurgy*, V.41, 15-31.

Peacock, D. P. S. 1977. Bricks and Tiles of the Classis Britannica: Petrology and Origin. *Britannia*, V.8, 235-248.

Peacock, D. P. S. 1982. *Pottery in the Roman World: An Ethnoarchaeological Approach.* Harlow: Longman.

Peacock, D. P. S. 1987. Iron Age and Roman Quern Production at Lodsworth, West Sussex. *The Antiquaries Journal*, V.67, 61-85.

Peacock, D. P. S. 2013. *The Stone of Life*. Southampton: The Highfield Press.

Pearson, A. F. 1999. Building Anderita: Late Roman Coastal Defences and the Construction of the Saxon Shore Fort at Pevensey. *Oxford Journal of Archaeology*, V.18 (1), 95-117.

Pearson, A. F. 2002a. *The Roman Shore Forts.* Stroud: Tempus Publishing.

Pearson, A. F. 2002b. Stone Supply to the Saxon Shore Forts at Reculver, Richborough, Dover and Lympne. *Archaeologia Cantiana*, V.122, 197-222.

Pearson, A. F. 2003. *The Construction of the Saxon Shore Forts*. Oxford: BAR Publishing.

Pearson, A. F. 2006. *The work of giants. Stone quarrying in Roman Britain.* Stroud: Tempus Publishing.

Perring, D. 1991. *Roman London.* London: Routledge.

Perring, D. 2009. *The Roman House in Britain.* London: Routledge.

Perring, D. and Roskams, S. 1991. *Early Development of Roman London West of the Walbrook.* York: CBA Research Report No 70.

Petrikovits, H von. 1975. *Die Innenbauten römischen Legionslager während der Prinzipatszeit.* Warsaw: Westdeutscher Verlag.

Pettitt, J. 1973. Reports of District Teams: Buxted. *Wealden Iron – Bulletin of the Wealden Iron Research Group*, V.1, 13.

Pettitt, P. and White, M. 2012. *The British Palaeolithic.* London: Routledge.

Philp, B. 1968. The Roman Cemetery at Keston. *Kent Archaeological Review*, Issue 11, 10.

Philp, B. 1972. The Discovery and Preservation of the Roman Villa at Horton Kirby. (Online). Available from: http://cka.moon-demon.co.uk/KAR030/KAR030_Horton.htm [Accessed 5th June 2015].

Philp, B. 1981. *The Excavations of the Roman Forts of the Classis Britannica in Dover 1970-1977*. Dover: Kent Archaeological Rescue Unit.

Philp, B. 1982. Romney Marsh and the Roman Fort at Lympne. *Kent Archaeological Review*, Issue 68, 175-191.

Philp, B. 1984. *Excavations in the Darent Valley, Kent.* Dover: Kent Archaeological Rescue Unit.

Philp, B. 1989. *The Roman House with Bacchic Murals at Dover.* Dover: Kent Archaeological Rescue Unit.

Philp, B. 1990. Excavations on the Roman Villa at Folkestone 1989. (Online). Available from: http://www.cka.moon-demon.co.uk/KAR099/KAR099_Folkestone.htm [Accessed 14th June 2014].

Philp, B. 1994. *The Iron Age and Romano-British Site at Lenham, Kent.* Dover: Kent Archaeological Rescue Unit.

Philp, B. 2012. *The Discovery and Excavation of the Roman Shore-Fort at Dover, Kent.* Dover: Kent Archaeological Rescue Unit.

Philp, B. 2014. *Discoveries and Excavations Across Kent, 1970 - 2014.* Dover: Kent Archaeological Rescue Unit.

Philp, B. and Chenery, M. 1997. *A Roman Site at Springhead (Vagniacae) Near Gravesend.* Dover: Kent Archaeological Rescue Unit.

Piggott, C. M. 1937, 231. East Belgic Pottery from Sussex. *Sussex Notes and Queries*, V.6, 231-232.

Pirie, E. 1957. Researches and Discoveries in Kent. *Archaeologia Cantiana*, V.71, 228 - 229.

Pitassi, M. 2012. *The Roman Navy.* Barnsley: Seaforth.

Pitts, M. 2014. London Eagle Watched Over Roman Tomb. *British Archaeology*, January/ February 2014, Issue 134, 8.

Plouviez, J: 1995. A hole in the distribution map: the characteristic of small towns in Suffolk. In: Brown, A.E. ed. *Roman Small Towns in Eastern England and beyond.* Oxford: Oxbow Books, 69-80.

Pollard, N. 2000. *Soldiers, Cities and Civilians in Roman Syria.* Ann Arbor: University of Michigan Press.

Pollard, R. J. 1988. *The Pottery of Roman Kent.* Maidstone: Kent Archaeological Society Monograph 4.

Poste, B. 1849. Proceedings. *Journal of the British Archaeological Association*, V.4, 65-81.

Poste, B. 1858. Observations on the site of Ancient Roman Maidstone. *Archaeologia Cantiana*, V.1, 154-175.

Poste, B. 1859. Ancient Rochester, or Durobrivae, Viewed as the Site of a Roman Camp or Station. *Archaeologia Cantiana*, V.2, 65–74.

Potter, T. W., Johns, C., Hall, D., Hassall, M. and Shotter, D. 1981. The Roman Occupation of the Central Fenland. *Britannia*, V.12, 79-133.

Potter, T. W. and Jackson, R. P. J. 1982. The Roman Site of Stonea, Cambs. *Antiquity*, V.217, 111-120.

Potter, T.W. 1996. Discussion and Conclusions. In: Jackson, R.P.J. and Potter, T.W. eds. *Excavations at Stonea, Cambridgeshire, 1980-1985.* London: British Museum Press.

Priestley-Bell, G. 2004. *The Excavation of Late Iron Age and Early Roman Remains on the Proposed Site of the A228 Roundabout and Services Bridge, (New Medway Works and Quarry, Holborough, Kent).* Unpublished: Archaeology South East, 2004)

Pritchard, F. A. 1986. Ornamental Stonework from Roman London. *Britannia*, V.17, 169-189.

Proctor, J. 2012. Roman Faverdale, A Frontier Trading Settlement. *Current Archaeology*, Issue 273, V.23, 20-25.

Pryor, F. 2004. *Britain AD.* Glasgow: Harper Collins.

Rackham, O. 1997. *The Illustrated History of the Countryside.* London: Orion Publishing.

Rady, J. 1992. *Archaeological Investigation at Teston, Maidstone, Kent.* Unpublished: Canterbury Archaeological Trust.

Rady, J., and Shand, G. 2004. *An Archaeological Evaluation of the Former ADT Building Site, Florence Road, Maidstone.* Unpublished: Canterbury Archaeological Trust.

Rae, A. and Rae, V. 1974. The Roman Fort at Cramond, Edinburgh. Excavations 1954-1966. Britannia. V5, 163-224.

Reece, R. 1980. Town and country: the end of Roman Britain. *World Archaeology*, V.12/1, 77-92.

Reece, R. 1981. The Third century, crisis or change? In: King, A. and Henig, M. eds. *The Roman West in the Third century.* Oxford: BAR Publishing.

Riccoboni, P. 2008. *An Archaeological Investigation at 98-99 Florence Road.* Archaeology South East, 2004).

Richardson, A. 2015. Up on the Cliffs. *Kent Archaeological Society Newsletter*, Issue 102, 16 - 21.

Rickman, G. 1980. *The Corn Supply of Ancient Rome.* Oxford: Oxford University Press.

Riddler, I. 2010. Querns. In: Bennett, P., Ridler, I. and Sparey-Green, C. eds. *The Roman Watermills and Settlement at Ickham, Kent.* Canterbury: Canterbury Archaeology Trust, 251 - 254.

Riddler, I. and Trevarthen, M. 2006. The Prehistoric, Roman and Anglo-Saxon Funerary Landscape, Saltwood, Kent. Rochester: CTRL Integrated Site Report Series.

Rigold, S. E. 1969. The Roman Haven of Dover. *The Archaeological Journal*, Issue 126, 79-100.

Rigold, S. E. 1972. Roman Folkestone Reconsidered. *Archaeologia Cantiana,* V.87, 31-41.

Roach-Smith, C. 1842. Account of Various Roman Remains Found in a Field Called the Slade in the Parish of Boughton Monchelsea, in Kent. *Archaeologia*, V.29, 414-419.

Roach-Smith, C. 1844a. Roman London. *The Archaeological Journal*, V.1, 108-117.

Roach-Smith, C. 1844b. Roman London. *The Archaeological Journal*, V.1, 108-117.

Roach-Smith, C. 1844c. Proceedings. *The Archaeological Journal*, V.1, 164.

Roach-Smith, C. 1848. *Collectanea antiqua: etchings and notices of ancient remains, illustrative habits, customs, and history of past ages.* London: J.R. Smith.

Roach-Smith, C. 1850. *The Antiquities of Richborough, Reculver, and Lymne.* London: J.R. Smith.

Roach-Smith, C. 1876. On a Roman villa Near Maidstone. *Archaeologia Cantiana*, V.10, 163-172.

Robertson, A. S. 1974. Romano-British Coin Hoards: Their Numismatic, Archaeological and Historical Significance. In: Casey, J. and Reece, R. eds. *Coins and the Archaeologist.* Oxford: BAR Publishing, 12-36.

Robertson, S. 1876. Medieval Folkestone. *Archaeologia Cantiana*, V.10, civ.

Robertson, S. 1883. Traces of Roman Occupation in and Near Maidstone. *Archaeologia Cantiana,* V.15, 68–88.

Rock, J. 1879. Ancient Cinderheaps in East Sussex. *Sussex Archaeological Collections*, V.29, 176-180.

Rodgerds, N. and Dodge, H. 2009. *The History and Conquests of Ancient Rome*. London: Hermes House.

Rogers, A. 2011. *Late Roman Towns in Britain. Rethinking Change and Decline*. Cambridge: Cambridge University Press.

Rowsome, P. 1996. The Billingsgate Roman House and Bath – Conservation and Assessment. *London Archaeologist*, V.7/16, 421-422.

Rowsome, P. 1999. The Huggin Hill baths and bathing in London: barometer of the town's changing circumstances? In: DeLaine, J. and Johnston, d. eds. *Roman Baths and Bathing*. Portsmouth: Journal of Roman Archaeology Supplementary Series Number 37, 263-277.

Rudling, D. 16th November 2013. Roman Period Settlement and Land-use in the Sussex Ouse - Presentation at CBA South East Conference on Landscapes of South Eastern Britain During the Roman Period.

Rudling, D., Cartwright, C., Swift, G., Foster, S., Shepherd, J., Hinton, P. and Tebbutt, F. 1986. The Excavation of a Roman Tilery on Great Cansiron Farm, East Sussex. *Britannia*, V.17, 191-230.

Rule, M. and Monaghan, J. 1993. *A Gallo-Roman Trading Vessel from Guernsey: The Excavation and Recovery of a Third century Shipwreck*. St Peter Port: Guernsey Museum Monograph 5.

Russel, A. 5th October 2002. Archaeology Unit Manager, Southampton City Council Archaeology Unit, (Speech at Museum of London Conference on *Classis Britannica*, accessed at http://www.classis-britannica.co.uk/sml/index.htm)

Russel, A. and Elliott, G. L. 2013. *Survey of Roman Structures in the River Itchen between Clausentum and St Denys, Southampton*. Unpublished: Southampton Archaeology Unit.

Russell, B. 2013a. *The Economics of the Roman Stone Trade*. Oxford: Oxford University Press.

Russell, B. 2013b. Gazetteer of Stone Quarries in the Roman World. (Online). Available from: www.romaneconomy.ox.ac.uk) [Accessed 12th January 2015].

Russell, C. and Staveley, D. 2012. A Geophysical Survey at Great Cansiron Farm, Butcherfield Lane, Hartfield, East Sussex. Unpublished: Chris Butler Archaeological Services Ltd.

Salway, P. 1970. The Roman Fenland. In: *The Fenland in Roman times: studies of a major area of peasant colonization with a gazetteer covering all known sites and finds*. Phillips, C.W. ed. London: Royal Geographic Society, Research Series No.5, Pp 1-21.

Salway, P. 1981. *Roman Britain*. Oxford: Oxford University Press.

Saynor, J. 1974. Roman Cremation Group From East Farleigh. (Online): Available from: http://cka.moon-demon.co.uk/KAR035/KAR035_Farleigh.htm [Accessed 15th July 2013].

Schäfer, A. and Trier, M. 2013. Cologne – Revealing a Roman Gateway to the Rhine. *Current World Archaeology*, Issue 59, 32-36.

Scott, A. 1972. Field Notes. *Wealden Iron – Bulletin of the Wealden Iron Research Group*, V.1, 29.

Scott, E. 1993. *A Gazetteer of Roman Villas in Britain*. Leicester: Leicester University Archaeological Research Centre.

Selkirk, A. 2011. Folkestone: Roman Villa or Iron Age Oppidum? *Current Archaeology*, Issue 262, V.22, 23-30.

Selkirk, R. 1983. *The Piercebridge Formula*. Cambridge: Patrick Stephens Limited.

Selkirk, R. 1995. *On the Trail of the Legions*. Ipswich: Anglia Publishing.

Selkirk, R. 2001. *Chester-le-Street and Its Place in History*. Durham: CASDEC Ltd.

Severn, J. 1975. *The Teston Story*. Teston: Rufus Fay Publications.

Shaffrey, R. 2006. *Grinding and Milling. A study of Romano-British rotary querns and millstones made from Old Red Sandstone*. Oxford: BAR Publishing.

Shaffrey, R. 2015. Intensive Milling Practices in the Romano-British Landscape of Southern England: Using Newly Established criteria for Distinguishing Millstones from Rotary Querns. *Britannia*, 46 (October 2015), 55-92.

Shaffrey, R. and Allen, J. R. L. 2014. A Complete Whetstone of Wealden Lithology from the Roman Site at Tackley, Oxfordshire. *Britannia*, V.45, 288-293.

Sheldon, H. 2010. Enclosing Londinium. *London and Middlesex Archaeological Society Transactions*. V.61, 227-235.

Skelton, S. 2010. *UK Vineyards Guide*. London: Stephen Skelton.

Smart, J. G. O., Bisson, G. and Worssam, B. C. 1975. *Geology of the Country around Canterbury and Folkestone - British Geological Survey Memoir*. Nottingham: British Geological Survey.

Smith, A. 2013. Roman Britain…As You've Never Seen It Before. *British Archaeology*, Issue 132, 20-23.

Smith, J. 1839. *The Topography of Maidstone and its Environs*. Maidstone: J. Smith Publisher.

Smythe, C. T. 1883. A Walled Roman Cemetery in Joy Wood, Lockham, Near Maidstone. *Archaeologia Cantiana*, V.15, 81–88.

Smythe, J. A. 1937. Ancient Sussex Iron Blooms. *Transactions of the Newcomen Society*, V.17, 197-203.

Southern, P. 2013. *Roman Britain*. Stroud: Amberley Publishing.

Sowan, P. W. 1977. Reigate Stone in Kent. *Kent Archaeological Review*. (Online). Available from: http://cka.moon-demon.co.uk/KAR048/KAR048_Reigate.htm [Accessed 10th February 2014].

Spain, R. J. 1984. Romano-British Watermills. *Archaeologia Cantiana*, V.100, 101-128.

Spain, R. J. and Riddler, I. 2010. Millstones. In: Bennett, P., Ridler, I. and Sparey-Green, C. eds. *The Roman Watermills and Settlement at Ickham, Kent*. Canterbury: Canterbury Archaeology Trust, 277-283.

Spencer, D. 2013. *The Dean Street Quarry - BA Dissertation*. Unpublished: University of Kent.

Starr, C. G. 1941. *The Roman Imperial Navy 31BC – AD 324*. New York: Cornell University Press.

Stathakopoulos, C. 2007. *Famine and Pestilence in the Late Roman and Early Byzantine Empire*. London: Routledge.

Staveley, D. 2012. *Latest Results: Teston*. (Online). Available from: http://geophyswithsnuffler.blogspot.co.uk/2012/07/latest-results-teston.html [Accessed 2nd Febuary 2016].

Staveley, D. 16th November 2013. The Sussex Roman Road Network - Presentation at CBA South East Conference on Landscapes of South Eastern Britain During the Roman Period.

Staveley, D. 2014. *Digging Up The Geophysics at Oaklands Park*. (Online). Available from: http://geophyswithsnuffler.blogspot.co.uk/2014/05/digging-up-geophysics-at-oaklands-park.html [Accessed 28th November 2015].

Stevens, S. 2014. Archaeological Investigations at Maidstone Hospital. Hermitage Lane, Barming. *Archaeologia Cantiana*, V.134, 141-152.

Straker, E. 1928. Roman Ironworks Near East Grinstead. *Sussex Archaeologocal Connections*. V. 69, 183-185.

Straker, E. 1931. *Wealden Iron*. London: Bell.

Straker, E. 1934. Wealden Iron Bloomery. *Archaeologia Cantiana*, Vol 46, 207.

Straker, E. and Lucas, B. H. 1938. A Romano-British Bloomery in East Sussex. *Sussex Archaeological Collections*, V.79, 224-236.

Straker, E. and Mason, R. T. 1939. A New Roman Bloomery Site. *Sussex Notes and Queries*, V.7, 153-154.

Strong, D. E. 1968. The Monument. In: Cunliffe, B. ed. 5th *Report on the Excavations on the Roman Fort at Richborough, Kent*. Oxford: Oxford University Press, 40-73.

Stuart-Hutcheson, A. 2009. *An analysis of the East Sussex hoard; the significance of the hoard for understanding the Roman Weald - BA Dissertation*. Unpublished, Kings College London.

Stuart-Hutcheson, A. 2012. *The High Weald Roman Coin Hoard*. (Online). Available from: http://brightonmuseums.org.uk/discover/2012/07/13/the-high-weald-roman-coin-hoard/ [Accessed 12th August 2013].

Stukeley, W. 1776 (2010 edition). *Itinerarium Curiosum*. Andover: Gale ECCO.

Swift, G. 1982. Minepits Surveys. *Wealden Iron – Bulletin of the Wealden Iron Research Group*, V.2, 15-21.

Symonds, M. 2013. Roman Track and Ore. *Current Archaeology*, Issue 279, V.24, 6-7.

Symonds, M. 2014. Emerson's Green: Live and Let Dye. *Current Archaeology*, Issue 287, V.25, 8.

Symonds, R.P. and Wade, S.M. 1989. A Remarkable Jar Found Inside an Amphora Cremation Chamber at Colchester. *Journal of Roman Pottery Studies*, V.2, 85.

Tatton-Brown, T. 1996. *Kent Churches - Architectural & Historical Information*. (Online). Available from: http://www.kentarchaeology.org.uk/01/03/BIG.htm [Accessed 15th January 2015].

Tatton-Brown, T. 2014. Gossenstein: Discovering an unexpected source of English Cathedral Columns. *Current Archaeology*, Issue 289, V.25, 38-41.

Taylor, C. C. 1982. *'The Nature of Romano-British Settlement'*, In: Miles, D. ed. *Romano-British Countryside*. Oxford: BAR Publishing, 1-15.

Taylor, J. 1994. The idea of the villa. Reassessing villa development in south-east Britain. In: Roymans, N. and Derks, T. ed. *Villa Landscapes in the Roman North*. Amsterdam: Amsterdam University Press, 179-194.

Taylor, J. 2000. Review of 'Stonea in its Fenland context: moving beyond an Imperial Estate.' *Journal of Roman Archaeology*, V.13, 647–658.

Taylor, J. 2007. *An Atlas of Roman Rural Settlement in England*. York: CBA Research Report No 151.

Taylor, J. 2013. Encountering Romanitas: Characterising the Role of Agricultural Communities in Roman Britain. *Britannia*, V.44, 171-190.

Taylor, M. V. 1932. Country Houses and Other Buildings. In: Page, W. ed. *The Victoria History of the Counties of England – Kent*, London: St. Catherine Press, 102-127.

Taylor, M. V., Jessup, R.F., and Hawkes, C.F.C. 1932. Topographical Index. In: Page, W. ed. *The Victoria History of the Counties of England – Kent*. London: St. Catherine Press, 144-177.

Taylor, T. 1998. Beauport Park – Time Team Episode. http://www.channel4.com/programmes/time-team/on-demand/25721-007 [Accessed 25th November 2014].

Taylor-Wilson, R. 2002. *Excavations at Hunts House, Guy's Hospital. London Borough of Southwark.* London: Pre-Construct Archaeology Monograph 1.

Tebbutt, C.F. 1971. Blacklands, Cansiron Bloomery. *Wealden Iron – Bulletin of the Wealden Iron Research Group*, V.1, 11.

Tebbutt, C.F. 1972. A Roman Bloomery at Great Cansiron, Near Holtye, Sussex. *Sussex Archaeological Collections.* V.110. 10-13.

Tebbutt, C.F. 1973. A Roman Site at Howbourne Farm, Hadlow Down. *Sussex Archaeological Collections.* V.111. 115.

Tebbutt, C. F. 1976. Kitchenham Farm, Ashburnham, Hadlow Down. *Sussex Archaeological Collections.* V.114. 324.

Tebbutt, C. F. 1978a. Mid Sussex Water Company Pipelines. *Sussex Archaeological Collections.* V.116. 402-406.

Tebbutt, C. F. 1978b. Reports on Fieldwork . *Wealden Iron – Bulletin of the Wealden Iron Research Group*, V1, 6-15.

Tebbutt, C. F. 1979. A Roman Coin from Great Cansiron Bloomery. *Wealden Iron – Bulletin of the Wealden Iron Research Group*, V.1, 14-15.

Tebbutt, C. F. 1981a. Wealden Bloomery Smelting Furnaces. *Sussex Archaeological Collections.* V.119. 57-64.

Tebbutt, C. F. 1981b. Field Reports. *Wealden Iron – Bulletin of the Wealden Iron Research Group*, V.2, 20-23.

Tebbutt, M. and Tebbutt, C. F. 1982. Oldlands Roman Bloomery. *Wealden Iron – Bulletin of the Wealden Iron Research Group*, V.2, 12-15.

Temin, P. 2012. *The Roman Market Economy*. Princeton: Princeton University Press.

Tester, J. P. 1963. Research and Discoveries in Kent: A Decapitated Burial at Cuxton. *Archaeologia Cantiana*, V.78, 181-182.

Thompson, F. 1986. The Iron Age Hillfort of Oldbury, Kent. Excavations 1983-84. *The Antiquaries Journal.* V.66, 267-286.

Todd, M. 1970. The Small Towns of Roman Britain. London. *Britannia*, V.1, 114-130.

Todd, M. 1996. Ancient Mining on Mendip, Somerset: A Preliminary Report on Recent Work. *The Bulletin of the Peak District Mines Society*, V.13 No.2, 47-51.

Tomlin, R. S. O. 1996. A Five Acre Wood in Kent. In: Bird, J., Hassall, M.W.C. and Sheldon, H. eds. *Interpreting Roman London: Papers in Memory of Hugh Chapman.* Oxford: Oxbow Books, 209.

Tomlin, R. S. O. 2016. *Roman London's First Voices.* London: Museum of London Archaeology.

Trinder, B. 2013. *Britain's Industrial Revolution.* Lancaster: Carnegie Publishing.

Tyers, P. 1996. *Roman Pottery in Britain.* London. Batsford Ltd.

Veen, van der M., Livarda, A., and Alistair Hill. 2008. New Foods in Roman Britain – Dispersal and Social Access. *Environmental Archaeology.* V.13, 11-36.

Vincent, A. 2007. *Roman Roads of Kent.* Midhurst: Middleton Press.

Visser, R. M. 2009. Growing and Felling Romans: Theory and Evidence Related to Silvicultural Systems. In: Moore, A., Taylor, G., Girdwood, P., Harris, E. and Shipley, L. eds. *TRAC 09: Proceedings of the 19th Annual Theoretical Roman Archaeology Conference, Ann Arbor and Southampton.* Oxford: Oxbow, 11-22.

Wacher, A. 1965. Researchers and Discoveries in Kent. *Archaeologia Cantiana*, V.80, 228- 229.

Waddlegrove, A. C. and Waddlegrove, E. 1990. Archaeology and Research into Sea-Level during the Roman Era: Towards a Methodology based on Highest Astronomical Tide. *Britannia*, V.21, 253-266.

Wainwright, G. J. 1991. Exploring Our Past: Strategies for the Archaeological of England. London: Historic Buildings and Monuments Commission for Britain, 1991.

Wallace, L. 16th November 2013. A Newly Discovered Roman Villa in Bourne Park, Canterbury - Presentation at CBA South East Conference on Landscapes of South Eastern Britain During the Roman Period.

Walters, B. 1998. Huge Roman Quarry Found in North Wiltshire. *The Bulletin for the Association of Roman Archaeology*, Issue 6, 8-9.

Walton, P. 2008. The Finds from the River. In: Cool, H.E.M. and Mason, D.J.P. eds. *Roman Piercebridge, Excavations by D.W Harding and Peter Scott 1969 – 1981.* Durham: The Architectural and Archaeological Society of Durham and Northumberland, 286-293.

Ward, A. 1999. *An Archaeological Watching Brief At Tovil Mill, Tovil Hill, Maidstone.* Unpublished: Canterbury Archaeological Trust.

Ward, A. 2013. More Confusing Bits of the Archaeology of Rochester. *Kent Archaeological Review*, Issue 191, 49-57

Ward, A. 2014. More Archaeological Problems at Rochester Cathedral. *Kent Archaeological Review*, Issue 196, 204-215.

Ward Perkins, J. B. 1971. *'Quarrying in Antiquity: Technology, Tradition and Social Change (Proceedings of the British Academy, Volume LVII).* Oxford: Oxford University Press.

Warry, P. 2010. Legionary Tile Production in Britain. *Britannia*, V.41, 127-147.

Watson, S. 2004. Roman and Medieval Occupation at 8-10 Old Jewry, City of London. *London Archaeologist*, V.10, 264-270.

Watson, S. 8th March 2015. Excavations at Bloomberg London: New Discoveries Along the Walbrook. Presentation at the Royal Archaeological Institute.

Webster, D., Webster, H. and Perch, D. F. 1967. A Possible Vineyard of the Romano-British Period at North Thoresby, Lincolnshire. *Lincolnshire History and Archaeology*, Issue 2, 55-61.

Weekes, J. 2014. Interpretation of the Site. In: O'Shea, L. and Weeks, J. eds. *Evidence of a Distinct Focus of Romano-British Settlement at Maidstone? Excavations at Church Street 2011-12. Archaeologia Cantiana,* V.135, 147-150.

Weekes, R. 2012. MAAG Lead Scroll. Kent Archaeological Society Newsletter, Issue 94, 4.

Welch, M. 2007. Anglo-Saxon Kent to AD 800. In: Williams, H. ed. *The Archaeology of Kent to 800AD.* Woodbridge: The Boyden Press and Kent County Council, 187-250.

Wenban-Smith, F. 2007. The Paleolithic Archaeology of Kent. In: Williams, H. ed. *The Archaeology of Kent to 800AD.* Woodbridge: The Boyden Press and Kent County Council, 25-66.

Wheeler, R. E. M. 1932. Romano-British Remains – Towns. In: Page, W. ed. *The Victoria History of the Counties of England – Kent.* London: St. Catherine Press, 60-101.

Whitby, M. 2002. *Rome at War AD 293 – 696.* Oxford: Osprey Publishing.

White, R. 2007. *Britannia Prima – Britain's Last Roman Province.* Stroud: Tempus.

Wild, J. P. 2002. The Textile Industry of Roman Britain. *Britannia*, V.33, 1-42.

Wilkinson, P. 2005. *An Archaeological Investigation of the Roman Aisled Stone Building at Hog Brook, Deerton Street, Faversham, Kent.* Unpublished: Kent Archaeological Field School.

Wilkinson, P. 2006. *The Historical Development of the Port of Faversham, Kent 1580-1780.* Oxford: BAR Publishing.

Wilkinson, P. 2008. *The Interim Results of an Archaeological Investigation at Stone Chapel Field, Faversham, Kent.* Unpublished: Kent Archaeological Field School.

Wilkinson, P. 2012. *Excavation of Octagonal Roman Bath-House at Bax Farm, Teynham.* Unpublished: Kent Archaeological Field School.

Williams, J. H. 1971. Roman Building Materials in South East England. *Britannia.* V.2. 166-195.

Williams, T. 1993. *The Archaeology of Roman London, Volume 3: Public Buildings in the South West Quarter of Roman London.* York: CBA Research Report 88.

Williams, T. and Swain, H. 2008. The Population of Roman London. In: Clark, J., Cotton, J., Hall, J., Sherris, R., Swain, H. eds. *Londinium and beyond. Essays on Roman London and its hinterland for Harvey Sheldon.* York: Council for British Archaeology. 33-40.

Willis, S. 2005. *Samian Pottery, a Resource for the Study of Roman Britain and Beyond: The Results of the English Heritage Funded Samian Project. An e-monograph.* [e-publication]. Internet Archaeology No. 17.

Willis, S. 2007. Roman Towns, Roman Landscapes: The Cultural Terrain of Town and Country in the Roman Period. In: Fleming, a. and Hingley, R. eds. *Prehistoric and Roman Landscapes.* Oxford: Windgather Press. 143-164.

Willis, S. 2008. Review of 'An Imperial Possession: Britain in the Roman Empire'. *The American journal of Archaeology*, V.112.4 (Online) Available from: http://www.ajaonline.org/online-review-book/587 [Accessed 10th January 2017].

Willis, S. 2010. *Brief Interim Report on ongoing Archaeological works and Observation at Harp Wood, Pedlinge.* Unpublished: University of Kent (Report Submitted to Kent HER).

Willis, S. 2011. Samian Ware and Society in Roman Britain and Beyond. *Britannia*, V.42, 167-242.

Willis, S. 2012. Review of 'The Roman Roadside Settlement at Westhawk Farm, Ashford, Kent'. *Britannia*, V.43, 433-434.

Wilson, P. 2002. *Roman Catterick and its Hinterland Volume 1.* York: CBA.

Winbolt, S. E. 1925. *Roman Folkestone.* London: Methuen and Co.

Woodcock, A. 1998. *A Gazetteer of Prehistoric, Roman and Saxon sites in Romney Marsh and the Surrounding Area.* (Online). Available from: http://rmrt.org.uk/gazetter-of-prehistoric-roman-and-saxon-sites-in-romney-marsh-and-the-surrounding-area/ [Accessed 10th January 2012].

Woodiwiss, S. 1992. General Discussion. In: Woodiwiss, S. ed. *Iron Age and Late Roman Salt Production and the Medieval Town of Droitwich.* York: CBA Research Report No 81, 183-193.

Worcester, R. 2010. *History of Allington Castle, Kent.* Unpublished: Allington Castle.

Worcester, R. 17th April 2013. *Rochester's Pivotal Role in the XII Century and Why it Matters Today - Speech at Business Guild Dinner, Rochester Cathedral.* (Online). Available from: http://magnacarta800th.com/wp-content/uploads/2013/04/MC-Rochester-Cathedral-speech-17.4.13-website1.pdf [Accessed 15th August 2014].

Worsfold, B. 1931. Romden Place and its Restoration. *Archaeologia Cantiana*, V.43, 73-84.

Worssam, B. C. 1963. *Geology of the Country Around Maidstone.* Nottingham: British Geological Survey.

Worssam, B. 1995. The Geology of Wealden Iron. In: Cleere, H. and Crossley, H. eds. *The Iron Industry of The Weald.* Cardiff: Merton Priory Press, 1-30.

Worssam, B. and Tatton-Brown, T. 1993. Kentish Rag and Other Kent Building Stone. *Archaeologia Cantiana*, V.112, 93-126.

Wright, T. 1854. *Wanderings of an Antiquary*. London: J. B. Nichols.

Yates, D. 2004. Kent in the Bronze Age: Land, power and prestige c.1500-c.700BC. In: Lawson, T. and Killingray, D. eds. *An Historical Atlas of Kent.* Chichester: Phillimore, 13-15.

Yates, N. and Gibson, J. M. 1994. *Traffic and Politics – The Construction and Management of Rochester Bridge AD43 -1993.* Woodbridge: The Boydell Press.

Yorke, B. 1995. *Wessex in the Early Middle Ages.* Leicester: Leicester University Press.

Young, C. J. 1981. The Late Roman Water-Mill at Ickham, Kent, and the Saxon Shore. In: Detsicas, A. ed. *Collectanea Historica: Essays in Memory of Stuart Rigold.* Maidstone: Kent Archaeological Society, 32-39.

Zerbini, A. 2014. The Late Antique Economy: Regional Surveys. *Late Antique Archaeology*, V.10 (1), 41-60.

Appendices

Appendix A – Building Materials Sourced in Kent and the South East During the Roman Occupation

Here I detail the specific stone types extracted and utilised in Kent and the South East during the occupation, noting that what follows is not designed to be exhaustive given the amount of settlement across the region during the Roman occupation, but is designed to be a general reference when working with the three regional surveys in the text of the work above. I have chosen to define building materials as dressed stone, undressed stone, flagstones, rubble core, that ground to create mortar and other bonding materials, and that crushed to provide flooring or foundations for subsequent building activity. Dates for the examples of the various types of stone used have been given where available to provide a chronological reference point. The materials are listed as they appear, top to bottom (generally), in the nine geological formations detailed in Section 2.1 in Chapter 2 above.

Cement Stone and Septaria

Nodules of such material, originating in the London Clays of the north Kent coast (Allen and Fulford, 1999, 169), were used as wall facing material in the south and west walls of the later 3rd century AD Saxon Shore fort at Richborough and as part of the wall core filling at the early 3rd century AD Saxon Shore fort at Reculver, and was also exported north for use as a facing stone in the walls of the mid-3rd century AD East Anglian Saxon Shore fort at Walton Castle (Pearson, 2002b, 201), joining the similar but subtly different cement stone quarried more locally from around the Naze (Allen and Fulford, 1999, 169). Pearson (2006, 71) explains that where available, such stone would have been preferred as a building material when compared to local flints, given the comparative ease with which such facing stones could be shaped. Meanwhile, Allen and Fulford (2004, 25) have also identified local cement stone as a source of some of the tesserae discovered at the Eccles villa in the Medway Valley in its 1st century and 2nd century AD phases.

Thanet Sandstone

This soft, pale grey sandstone from the Thanet Beds was used in the north, west and south wall facings of the Saxon Shore fort at Richborough. The material would have been sourced from doggers found on the beaches at Pegwell Bay and further north at Herne Bay (Pearson, 2002b, 207).

Chalk

Locally quarried chalk from the North Downs and the east Kent coast around Deal (Lyne, 1994, 132) was extensively used as a building material during the Roman occupation, particularly in the eastern region covered by this research. The quarrying would have been on a much smaller scale than that detailed below for Greensand and ragstone in Kent, an analogy from elsewhere in occupied Britain being the occupation-period chalk quarry pits on Ridgeway Hill near Weymouth (later re-used as mass graves for decapitated Viking warriors) which were no more than 1.7m deep and 7m wide (Loe *et al.* 2014, 22). The quarried chalk was used in the occupation both as a primary building material in its own right as ashlar blocks, or crushed for wall infill, road-building, flooring, and the manufacture of mortar and plaster (Pearson, 2002b, 210). Examples of the former, despite Pearson (2006, 123) describing it as a poor building material (it being more appropriate as an interior facing), include its use with tufa (in alternately layered blocks) in the walls of the early 2nd century AD *Classis Britannica* fort in Dover (Philp, 1981, 20), in the bastions (possibly re-used) of the later 3rd century AD Saxon Shore fort on the same site, and additionally its use alone in the Mithraeum/ cellar at Wouldham/ Burham (Taylor, 1932, 109, see Section 4.1.4) where three incised chalk building blocks were found, they now being stored in Maidstone Museum.

Meanwhile, occupation-period lime kilns for the manufacture of various types of plaster (making use of the chalk) are found in a variety of places across Kent and the south east, for example outside the Saxon Shore fort at Richborough (Jessup, 1932, 128).

Flint

Flint, sourced from the North Downs and from beach deposits (for example the storm beaches on Thanet, Pearson, 2002b, 198), was used as a building material across the region during the occupation, often as a flint rubble core for major walls, examples being found at the Saxon Shore forts at Reculver and Richborough (Pearson, 2002a, 25). It was also extensively used as a facing stone set in a durable mortar, for example at the 2nd to late 4th century AD Crofton villa in Orpington (Taylor, 1932, 110), the early 3rd century AD 'Painted House' in Dover, the later Saxon Shore fort there (Philp, 1989, 31), and at a variety of sites in Canterbury (Jessup, 1932, 61).

Ferruginous Sandstone

Quarried from the Folkestone Beds on the east Kent coast, Ferruginous sandstone was used across the South East, for example forming the decorative band in the north wall of the Saxon Shore fort at Richborough (Pearson, 2002b, 205) and being used as a capping stone on the Severan land walls of London. Its notable red colour is derived from the iron-oxides which bind the grains of sand together.

Ironstone

In addition to its use as an iron-ore, this was also quarried locally as a building material along the foreshore of eastern Kent, being used as a material for building foundations and wall core filling at sites such as the late 1st century AD Villa 1 at East Cliff in Folkestone. Ironstone, named after its very high iron-oxide content, is created either by the same ferruginous process as the sandstone above, or by chemical replacement. Its use as a foundation material is based on its comparative hardness, for example when compared to Ferruginous Sandstones.

Greensand

Various types of greensand were quarried from the layers of the Lower Greensand Formation during the occupation, particularly along the east Kent coast, and were widely exported as a building material across southern and eastern Britain. Williams (1971, 172) said it is a much better building stone than chalk or flint as it is both workable (being capable of precise working into long blocks and slabs) and very durable. Prominent examples of its use locally include at the Kentish Saxon Shore forts at Reculver and Richborough (Pearson, 2002b, 202), and the late 1st century AD bath house at Beauport Park (Brodribb et al. 1988, 232), while further afield it is also found exported for use in the East Anglian early 3rd century AD Caister-on-Sea and later 3rd century AD Bradwell-on-Sea Saxon Shore forts (Allen and Fulford, 1999, 169). Dark green greensands, very heavy in glauconite, are also found used as significant-sized ashlar blocks in the jambs of the city gates of Roman Canterbury (notably Worthgate, Ridigate and Queningate), and as the foundation of the west gateway at the Saxon Shore fort in Richborough (Worssam and Tatton-Brown, 1993). At the opposite end of the spectrum in terms of size, dark-green tesserae heavy in glauconite have also been identified as a local greensand by Allen and Fulford (2004, 25) in the 1st century AD phase of the Eccles villa alongside the cement stone described above. Meanwhile, very hard Folkestone Beds greensand blocks, rounded and thus indicating their origins on the beaches of the eastern Kentish coast, are found as a facing stone in some parts of the later 3rd century AD walls of Roman Canterbury. The greensand would mostly have been quarried in the Folkestone (or nearby Eastbourne) area, either inland or from easily accessed beachfront coastal outcrops (as with the Folkestone Beds stone above).

Ragstone

This ubiquitous fine quality building stone, from the Hythe Beds in the Lower Greensand Formation, is distinct enough from other Lower Greensand materials to justify its own entry, particularly as it plays such an extensive part in the Medway Valley regional analysis of this research. It was utilised across the South East during the occupation in huge quantities, either as finely-worked dressed facing stone, in-fill rubble or to create hard road surfaces. The grey-green sandy and glauconitic limestone outcrops principally in the region of Maidstone (where the best quality stone is found), near Sevenoaks, and at the eastern extremity of the Greensand Ridge where the Hythe Beds outcrop in the cliffs near Folkestone and Hythe. Its uses are too numerous to mention here, though many instances are detailed in the narrative below. Hill and David (1995, 17) say that ragstone is still much in demand today, particularly for hammer-dressed walling. In that regard, since the Roman period it has been widely used, for example in the construction of the Westgate Tower in Canterbury and Maidstone Prison. It is currently being quarried at two sites, the principal one being Gallagher Group's Hermitage Quarry in Barming.

Tufa

Tufa is a variety of calcerous limestone, created by a secondary process involving the precipitation of carbonate minerals derived from ambient temperature bodies of water. It was used in a variety of forms as a building material during the occupation in Kent, predominantly in the east of the county, for example in many of the buildings and fortifications around Dover and Folkestone. Parfitt and Philp (1981, 176) and Allen and Fulford (1999, 169) all believe that much of the material was quarried in the valley of the River Dour (the original material deriving from the Hythe Beds), which flows through Dover. Specific examples of it being used as a primary building stone include the walls of the *Classis Britannica* fort in Dover (Philp, 1981, 20), the bastions of its later Saxon Shore successor, the facing of the nearby *pharos,* the north wall of the Saxon Shore fort at Richborough, in the walls of the 3rd century AD Saxon Shore fort at Lympne (only a small quantity and almost certainly re-used from an earlier structure), in the late 1st century AD large aisled building located at Hogs Brook near Faversham (Wilkinson, 2005, 4), in the walls of the large villa at Eccles in all of its phases (Williams, 1971, 174) and indeed similarly at Villa 1 at East Cliff in Folkestone according to Parfitt (2013, 41). He is particularly surprised by the latter, given the immediately available resources of greensand for building construction, this being a finer quality of building stone. He speculates that it reflects the scale of nearby tufa quarrying in the Dour Valley at the time of the villa's construction. Tufa's porosity and lightness also lent itself readily for use in complex vaulting structures, for example bathhouse roofs. It is tufa from the bathhouse of the 2nd century AD phase of the East Farleigh villa site that has been identified by the author re-used in the walls of St Mary's Church in the village. Meanwhile, tufa was also exported widely from Kent, for example providing dressed blocks for the East Anglian Saxon Shore fort at Walton Castle (Allen and Fulford, 1999, 169) which were later re-used in the nearby church of St Peter and Paul in Old Felixstowe, this reflecting the experience at St Mary's in East Farleigh and also All Saints at West Farleigh with their re-used Roman tufa. Finally, in addition to it being a sought after primary building stone, tufa rubble was also used as a foundation material or wall coring material during the occupation.

Grey Wealden Shale

Finely cut tiles of high quality Grey Wealden Shale have been found across Roman London and at the extensive 1st century villa AD at Fishbourne in Sussex for use in *Opus Sectile* tiled flooring (Pritchard, 1986, 185). Its fine quality is testified to by the other materials used in the tile floors where it is found, including Carrera and Purbeck marble. The grey limestone tiles, with a distinctive muddy texture, were found cut into equilateral and right angled triangles and seem to have been a feature of the early occupation, given *Opus Sectile* flooring fell out of use in Britain in the early 2nd century AD. Some of the tiles were later re-used however, and have been found in contexts as late as the 4th century AD. The material was quarried either in the Weald or, less likely, in the Medway area from deposits sitting within the Wealden Clays or the Wealden Beds.

Ashdown Sandstone

Allen and Fulford (1999, 179) detail that Ashdown Sandstone from the Ashdown Formation in the Hastings Beds was quarried on the eastern/ coastal edge of the Weald during the occupation, it being exported for use in the 3rd century AD East Anglian Saxon Shore fort at Brancaster.

Purbeck Limestone

Lott and Cameron (2005, 2) detail that inlier beds of this fine quality fossiliferous limestone were also exploited in Kent and the South East during the occupation. They are found in the Purbeck Group, most famous for producing the very high quality stone quarried in Dorset and called Purbeck Marble. Featuring large clasts within a very fine mud matrix, Purbeck Limestone was well known for its ability to take a very fine polish (best seen in the Dorset Purbeck Marble variety), and also for ease of working. It was also in demand because of the variations in colour naturally available in the seams of the material. The best example of its use during the occupation in the South East is at the Beauport Park bathhouse where it was used in the bathing complex itself (Brodribb et al. 1988, 234).

Clay for Bricks and Tile

This was sourced from a variety of layers within the Wealden Dome. As detailed above, Pearson (2002a, 80) says tiles and bricks made from Wealden Clays were extensively used across the region, for example in the Saxon Shore forts as bonding layers within the walls and of course extensively and region-wide for roofing. With regard to the latter, Williams (1971, 178) said that regionally made *tegula* and *imbrex* first start appearing in the county in the later 1st century AD, citing examples from Canterbury. Meanwhile Gault Clay, originating in a band sitting between the Upper and Lower Greensand Formations (see above), was also used, as evidenced by the occupation-period buildings of ragstone and Gault Clay bricks recorded at Sharfleet Creek on the Medway Estuary in 1885 (Jessup, 1932). Clay was also used as a primary building material for wall structures, particularly early in the occupation (Perring, 2009, 110) though its use continued throughout the Roman period. An example of later use is evidenced by the excavations at Newgate Street, London where 2nd century AD houses were found with walls constructed of cob, a mix of straw and clay (Perring and Roskams, 1991, 78). Clay for such building activity would have been quarried locally to its place of use. Finally, and specifically with regard to the Wealden *Classis Britannica* tile industry (see above), Fairlight Clay from the Ashdown Formation of the Hastings Beds was the preferred material for this use (Peacock, 1977, 237).

Sand and Hassock for Mortar, Opus Caementicium and Opus Signinum

Sand is accessible as a building material in various sections of the strata of the Wealden Dome. The concrete and cement production known to have taken place in the lower Medway Valley would have made extensive use of regionally available raw materials, for example the quarried sand and also hassock, a calcerous sandstone comprising quartz, clay, carbonate of lime and iron oxide and which is interbedded with the local ragstone in the Hythe Beds. Also utilised would have been the detritus from ragstone quarrying and locally ground chalk (see above), while gravels would also have been quarried, this being found interbedded with the sand or from beach deposits.

Wood

A casual glance at a land use map of Kent during the occupation reveals a county which was still almost half covered in woodland, despite the fact that one of the key indicators of the four centuries of the occupation is evidence of woodland clearance through pollen analysis (Dumayne-Peaty, 1998, 319). Indeed, we have one of our best phenomenological glimpses into occupied Britain in the context of Kentish woodland, in the form of an early 2nd century AD transactional wooden tablet (originally covered in wax and found in the Walbrook Valley in London) which details the sale of two hectares of woodland in the land of the *Cantiaci* from one Titus Valerius Silvinus to Lucius Julius Bellicus for forty *denarii* (Tomlin, 1996, 209). It is unclear whereabouts in Kent the exact location of this parcel of woodland actually was, though if it were in the north then a reasonable assumption is that it would be agro-forestry as part of the wider agricultural landscape, and if in the Weald to the south then coppicing to support the iron industry. Tomlin (1996, 210) suggests a third option, that it is a sacred site dating back to the origins of the *Cantiaci* prior to the occupation. Whatever the location and purpose, this case clearly illustrates the value of woodland and the worth attached to their legal registration. More broadly Lawson and Killingray (2004, 20) are clear about the importance of woodland exploitation during the occupation, although they also acknowledge the taphonomic challenges faced by the archaeologist in this regard, saying: *"Timber production was…very important although this industry has left no trace."*

In terms of the use of wood in the built environment, Visser (2009, 11) explains that while much commentary regarding buildings in occupied Britain focuses on stone-built structures, the majority would have been at least partially timber-built using posts, beams or planks, with Hanson (1996, 3) adding that this was particularly the case earlier in the occupation. In this regard Goodburn (1991, 182) was able to reconstruct the building techniques used and the final form of timber structures in London based on the excavations at Cannon Street carried out by the Museum of London. He says that the sequence of construction of such a building would have been as follows:

- Frame components on the building fitted together on site, using chalk or charcoal to label each piece.
- Base plates laid down, on clusters of supporting piles (as at the Walbrook site) if the ground was likely to be wet. These would be up to 6m long for an urban dwelling, matching in size the room to be located above.
- Partition plates laid down.
- Studs and corner posts set into the mortices in the base plate.
- Top plates laid on top of the corner posts.
- Diagonal brace elements fastened to the studs.
- Horizontal in-fill staves fitted into sloping recesses in the sides of the studs, with vertical rods or laths woven between them.
- Roof structure erected.
- Tie beams dovetailed over the top plates to prevent the walls from spreading under weight of the roof. The tie beams would also have supported the feet of the principal roof rafters. Betts (pers. comm. 24 October 2012) says that the roofs of later buildings in London tended to be steeper, necessitating the use of a lighter tile when tiling was used.
- Daub, usually in a Roman context made of brick earth, applied to the infill. The studs and the baseplates would be left exposed on the inside of the building.
- Finally, the whole exterior would be clad in planks of up to 35cm in width and 3cm in thickness. In a London context such cladding was usually of oak.

All of these findings were found replicated in the later excavation at the early 2nd century AD Courage Brewery site in Southwark of a Roman timbered waterfront warehouse (Brigham et al. 1995, 1). Meanwhile, and intriguingly, the timbers which formed the frame of the most subsubstantial building in the Cannon Street project had been re-used, they originally forming part of the 1st century AD land reclamation programme which revetted the eastern bank of the Walbrook Stream.

Specific examples of where such timber has been used as a primary building material during the occupation in Kent (and indeed anywhere, Goodburn, 182, 1991) are few and far between, but one can be found at the Ickham Roman watermills site. Here, the millhouse buildings and associated infrastructure were largely constructed of wood, a common practice with regard to millhouses in the north western provinces of the Empire as opposed to the Mediterranean where their counterparts were usually constructed of worked stone or brick (Bennett, Ridler and Sparey-Green, 2010, 63). Another is the late 1st century AD bakery at Springhead where flat foundations of flint, layered over with tiles, have been interpreted as being the fittings for a half-timbered building (Williams, 1971, 175). Meanwhile, further afield at the Southwark Courage Brewery site but useful by way of analogy, Perring (2009, 126) talks of the fine quality of the timber floor surfaces which he believes were common in occupied Britain in his interpretation of Brigham, Goodburn, Tyers and Dillon's excavation report, he saying:

> *"At the Courage Brewery site in Southwark (the) plank floor was laid over joists at 500mm intervals. The planks were 300mm and 450mm wide and the joists were dovetailed or lap-jointed into the sill beams. The planks were rebated into a central beam across the middle of the building."*

Context

The building materials detailed above are specifically those extracted in Kent and the South East. However some of the finer decorative materials evident in occupation-period buildings in the region were actually imported from much further afield. Examples would be the Tuscan Carrera marble used in the triumphal arch at Richborough, Ditrupa Limestone from the Valois-Soissons region of Gaul found as small square blocks in the Saxon Shore fort at Reculver and in the North Wall of the Saxon Shore fort at Richborough (the latter possibly re-used from the arch), the Bath Oolite columns found at Eccles (Williams, 1971, 181) and the Caen Stone imported from the Calvados region of Gaul which found its way into both Kentish buildings and those in London (Sowan, 1977, 1). Nevertheless, as set out above, clearly the vast majority of the building stone used in occupation-period Kent and the South East were both quarried and utilised locally.

Appendix B – Sources of Data for the *Classis Britannica*

The data used to interpret the activities of the *Classis Britannica* in this book is derived from the archaeological record, the historical record, scientific observation and analogy, together with supporting anecdotal evidence where appropriate.

In terms of the archaeological many vessels of all sizes dating to the Roman occupation period have been found and examined across north-western Europe. Meanwhile, we also have thousands of tiles and bricks found on both sides of the English Channel bearing the 'CLBR' stamp of the *Classis Britannica*. These are specifically detailed above in Section 3.6 in their own section rather than below in this appendix.

Data from the historical record comes in the form of primary sources and epigraphy. Over 100 ancient works mention Britain and a number of authors go into great detail, with some specifically highlighting the regional navy in action. The most important for this research are Julius Caesar himself with his *Gallic Wars* (1st century BC but useful to set the scene for pre-conquest Britain), Strabo with his *Geography* (late 1st century BC/ early 1st century AD, useful again for pre-conquest context), Cornelius Tacitus with his *Annals, Histories* and *Agricola* (late 1st century AD to early 2nd century AD, and hugely important for the conquest period), Suetonius with *The Twelve Caesars* (late 1st century AD to early 2nd century AD, and again relevant to the conquest period), Cassius Dio with his *Roman History* (late 2nd to early 3rd centuries AD, his extensive work covering most of the existence of the *Classis Britannica*), Herodian with his *History of the Roman Empire* (late 2nd century AD to mid 3rd century AD, and particularly important for the campaigns in Britain of Septimius Severus), and Ammianus Marcellinus (4th century AD, with his *Roman History* and useful for analogy). To these authors and their titles we can add the now anonymous *Historia Augusta*, a collection of biographies of emperors from AD 117 to AD 284 written in the later Empire. Each is quoted or referenced as relevant throughout the research above.

Next we can examine a number of official itineraries and codices which include occupation-period Britain when giving details of the main travel routes across the Empire, crucially including place names for towns and key sites such as fortresses, and the distances between them (and thus helping determine the locations of the ports and harbours through which the *Classis Britannica* operated). Firstly we have the *Tabula Peutingeriana*, a damaged 13th century AD reproduction of a Roman road map which only includes the South East corner of Britain (the remainder having been being lost). Specifically detailed on the map are the Kentish ports which played a major role in the activities of the *Classis Britannica*, namely Richborough (a multiple purpose site featuring a harbour, the monumental arch detailed above and later a key Saxon Shore fort), Dover and Lympne. Next we have the *Antonine Itinerary*, this surviving as a manuscript which details the 225 most important roads in the Empire and the key sites along them. This provides a huge amount of information about occupied Britain, including detail of all of the main civic centres and fortresses. Appearing to date from the reign of Caracalla (sole Emperor from AD 211 to AD 217), it therefore overlaps with the period of activity of the *Classis Britannica*. Next, though not strictly an itinerary, we have the *Ravenna Cosmography*, an early 8th century AD list of names of places in the Empire based on earlier sources and thus including 300 locations in Britain (three centuries after the latter had ceased to part of the crumbling western Empire). To this we can add the *Notitia Dignitatum,* a late Roman collection of the key officials and military formations in the Empire and their locations (mentioned above in Section 2.6.1). Though outside the period of chronological interest of this book the *Notitia* is useful by way of analogy. Finally we have the *Codex Theodosianus*, a compilation of all legal rulings since the time of Constantine 1 dating to the first half of the 5th century AD. Again outside the time of interest for our period of study, it is useful by way of analogy.

We now move on to the epigraphy which specifically mentions the *Classis Britannica*, a number of references surviving in the context of memorials detailing the careers or activities of *praefecti classis*, *trierarchi* (Captains) and crewmembers of the fleet. In the case of the former a good example is provided by an inscription from Ostia naming Q. Baienus Blassianus as the *praefectus* of the *Classis Britannica*, this set up around AD 160. Another, closer to home, comes in the form of an alter dedicated to Neptune which was found re-used in the walls of the later 3rd century AD Saxon Shore at Lympne (almost certainly originating from the original *Classis Britannica* fort on the site). The epigraphy on the alter says that it was set up by Admiral of the British fleet Lucius Aufidius Pantera and has been dated to the mid-2nd century AD. Further, a Marcus Maenius Agrippa is detailed on an inscription from Camerinum in Umbria which states that he commanded the *Classis Britannica*, again in the mid 2nd century AD and with him likely being the successor to Pantera commanding the regional fleet. He is also mentioned on an alter inscription to 'Jupiter Best and Greatest' found at Maryport (Roman Alauna) in Cumbria where he is also listed as being the overall Procurator of the whole province, possibly occupying the two roles at the same time (see Section 2.6.2 above). Next we have Sextus Flavius Quietus on whose tombstone in Rome is detailed the fact that he became the commander of the British fleet, again in the mid 2nd century AD. Meanwhile an inscription from Celeia in Noricum details that one Titus Varius Priscus was also the *praefectus* of the *Classis Britannica*, though no date is available here, while finally for Admirals we also have an unknown individual who is detailed in a partial inscription from Rome as having commanded four regional fleets - the *Classis Britannica, Classis Germanica, Classis Moesica, Classis Pannonica.* This latter is particularly interesting as it is not clear if the reference is to consecutive commands or to an unusual combination of all four. This may refer to the state of affairs after the defeat of usurping British governor Clodius Albinus by Septimius Severus in AD 197. In this context the *Classis Britannica* would almost certainly have supported Albinus given it would have been needed to transfer his army across the English Channel to northern Gaul. We know that following his defeat the whole military in Britain was 'reformed' by the new governor and military commissioners were sent to the province by Severus. It may well be that while the 'reforms' were taking place a single commander could have controlled the British, Rhine and Danube fleets. One final mention here in terms of senior officer epigraphy should go to Olus Cordius Candidusa, a transport officer who is referenced to on an alter found in a *Classis Britannica* context at Dover (though note the epigraphy does not specifically mention the regional fleet, Henig, 1995, 69).

In terms of references to *trierarchus* of the *Classis Britannica*, we have five, the first four on gravestones at the regional fleet's headquarters in Boulogne. Firstly there is Quintus Arrenius Verecundo, then Tiberius Claudius Seleucus, followed by P. Graecius Tertinus (in the context of his sons memorial) and finally B. Domitianus (again in the context of a memorial to his son). Finally we have the North African Saturninus, known to have been a *trierarchus* of the *Classis Britannica* from an inscription found at Arles in southern France. This is a particularly important piece of epigraphy as it is dated to between AD 244 and AD 249 and is the last reference known to the *Classis Britannica*.

To conclude the personalised epigraphic evidence we have one Didio, marine of the *Classis Britannica* who is known from his gravestone at Boulogne, and finally Demetrius, another marine of the regional fleet also buried at Boulogne where a fragment of his gravestone has been found. The former originated in Thrace, while the latter came from Syria.

Epigraphy regarding the *Classis Britannica* not specific to an individual is also found in Britain. This includes the well-known inscription on a stone block in the portico of the granary of the Roman fort at Benwell, this referencing a *Vexillatio Classis Britannicae* (RIB1340, Breeze and Dobson, 2000, 66). Such inscriptions have traditionally been interpreted as an indication that the detachments actually built the section of wall or structure rather than they being permanently based there fulfilling a military mission. Additionally, the regional fleet is mentioned in the context of a *Cohors 1 Aelia Classica* (First Cohort of the Hadrianic Fleet) dating to AD 146 on a list of units detailed on a bronze discharge diploma from Chesters fort on Hadrian's Wall. This appears to be the same unit mentioned in another diploma dated to AD 158 found in three pieces outside the Roman Fort at Ravenglass, and on a lead seal from the same location.

Of the other forms of data, scientific observation is threaded throughout the whole narrative. Analogy is similarly extensively used, for example in the comparison of the experiences of the *Classis Germanica* detailed in Section 5.1.4. Finally, anecdote too is used extensively throughout this work with regard to the regional fleet and its activities where appropriate.

Appendix C – Roman Maritime Technology

In Chapter 4 above I considered whether hydraulic river infrastructure was used to facilitate navigation upriver of the tidal reach in the Medway Valley to support the intense ragstone quarrying industry located there (and the agriculture and settlement which supported it), while later in Chapter 5 I discussed the role played by the *Classis Britannica* in the ragstone quarrying industry there during the occupation. Both of these considerations and discussions were heavily influenced by Roman riverine maritime technology in the form of the ships and boats which would have facilitated the industry, agriculture and settlement. To that end I discuss here the specifics of such technology given it is threaded through the entirety of the above work.

McGrail (2015, 123) and Pitassi (2012, 23) argue that in north-western Europe during the Roman period there were two distinct ship building traditions, with vessels built either in a Mediterranean style or a Romano-Celtic style. In the case of the former the key distinction was the use of locked mortise and tenon plank fastenings. Many of these ships would have been military in nature to support campaigning on land and for dealing with pirates, usually in the form of Mediterranean-style *liburnae* biremes (war galleys, in Britain and northern Gaul these operating under the command of the *Classis Britannica* until its demise in the mid-3rd century AD). Lyne's (1996) analysis of Roman ship fittings found at Richborough has highlighted the preponderance of these *liburnae* there, he using data from the 1922-1938 excavations to show that such vessels were the most common type present, not surprising given the *Classis Britannica* presence earlier and the Saxon Shore fort later. The best-known example of a Mediterranean-style vessel from occupied Britain however is the County Hall ship discovered in London in 1910, interestingly built locally using oak from the South East of England. Dated to the late 3rd/ early 4th centuries AD (Marsden, 1994, 128), this ship is not thought to have been military in nature, with McGrail (2015, 123) arguing that it was a ferry used between the capital and the Continent. Other vessels built in the Mediterranean-style known from across north west Europe from this period include a galley found at the Roman fort at Laurum in The Netherlands, (Lendering, 2012, 1), a 1st/ 2nd century AD boat excavated at the Roman fort at Vechten in The Netherlands in 1893, two 1st/ 2nd century AD boats excavated at the Roman fort at Zwammerdam in The Netherlands in 1968-1971, and finally two riverboats found at Oberstimm in central Germany (McGrail, 2015, 123).

Far more common in north-western Europe however, and of greater relevance to this research given their mostly commercial usage, are vessels of Romano-Celtic design. Milne (2000, 131) explains that they are so called because of their similarity to the ships of the *Veneti* described by Caesar. Identified through some 30 wrecks found in the Severn Estuary, the Thames, the Channel Islands, the Schelde/ Meuse/ Rhine Delta, the Rhine at Zanten and at Mainz (McGrail, 2015, 124), vessels of this design broadly have the following features:

- A framing of closely spaced, large timbers with half-frames spanning the sides and bottom, and with a floor covering the bilges and bottom (the individual timbers not being fastened together).
- Planking which is flush-laid and fastened to the frame with large iron nails.
- Caulking within the plank seam using macerated twigs, moss or twisted fibre.
- Where a mast is used, the mast step being well forward.

McGrail (2015, 125) believes that within the Romano-Celtic ship building tradition there are two distinct groupings which he styles Type A and Type B. The former are keel-less and flat bottomed, designed for use on canals, lakes and rivers. Though some types had sails, most examples used a towing post set forward and are best illustrated by the numerous examples of *codicaria* river barges found in the archaeological record and in epigraphy across the region. Such vessels are heavily referenced by Ausonius (Mosella, 5) who speaks of the use of hawsers by the men towing them from the riverbanks. In a Medway Valley context, in unpublished research Milne provides insight into the operations of such towing-mast fitted barges, he saying (pers. comm. 08 March 2011):

> *"If you go against the current you will need to use a tow path which has to be kept clear. The tow path can be either side of the river though it seems in the case of the Medway it was is on the opposite bank to the quarries and associated industry which would make sense."*

Analogously we can get phenomenological insight into riverine barge operations on the Medway during the occupation by examining the experiences of later such activity in the 18th and 19th centuries once the Medway Navigation Company had opened up the navigation of the river. The same towing bank was used as during the Roman period, with the barges being pulled by gangs of men, called 'halers' or 'hufflers', as opposed to barge horses. Generally, once a vessel was underway and there was little or no current, a single person could reasonably be able to pull a 50 tonne barge load, though it was normal practice at that time for a number of men to be employed in this regard. For context, in this period it took a barge being pulled in this way 12 hours to get to Maidstone from Tonbridge. Maintenance of the bank was essential here to ensure the commercial success of the barging operations, this being carried out at the same time as riverbed maintenance to ensure the river remained fully navigable (Ellis Jones, 2012, 5). The experiences of those operating the barges during the 18th and 19th centuries seems to have been broadly similar to that of their Roman forebears, with Selkirk (1983, 83) detailing teams comprising a helmsman and four *codicarii infra pontem sub(licium)* (Roman 'hufflers') operating on towpaths along the River Tiber, taking barges from Ostia to Rome.

Other common examples of Type A vessels would have been flat-bottomed varieties of the ubiquitous smaller monoreme *myoparo* and *scapha* (types of cutter and skiff, Mason, 2003, 142). Such vessels are memorably described by Ausonius (Mosella, 19), he saying:

> *"What bright regattas charm us when, mid-stream, oar-driven skiffs engage in mimic war...the exultant captains move lightly on stern or prow, the youthful crew racing across the water...the picture of these painted prows, and crews of laughing lads, reflected in the river."*

Type A vessels found in the archaeological record include further examples at Laurum to join the galleys detailed above (Lendering, 2012, 1), a range of flat-bottomed river barges found during excavations of the harbour at Cologne (*Colonia Claudia Ara Agrippinensium,* Schäfer and Trier, 2013, 36), the Bevaix Boat from Lake Neuchatel in Switzerland, and flat-bottomed barges found at Pommeroeul in Belgium (Campbell, 2011, 146) on a tributary of the River Haine. The largest of these vessels had a central gangplank and is similar in size to a *codicaria,* while the smallest was skiff sized and likely a *scapha.* One of the vessels is now on display at the Gallo-Roman Museum in Ath.

Type B vessels are sea-going, though often capable of riverine use also, and have a full-bodied hull form with firm bilge, posts and a plank keel. They were propelled by a sail in estuaries and when at sea, and a number of examples have been found in Britain dating to the occupation. London provides two excellent candidates in the New Guys House boat found in 1958 and the larger Blackfriars 1 vessel excavated by Peter Marsden in 1962 (and subsequently used by Pearson in 4.3 above for his boatload estimations regarding Saxon Shore fort construction). The Blackfriars 1 ship is of particular importance here given it was found to be carrying 26 tonnes of Kentish ragstone, this being petrologically identified as originating in the Medway Valley (Spencer, 2013, 31). The Blackfriars 1 vessel was 14m in length and 6.5m wide, with a shallow draught of 1.5m and a maximum speed of around 7 knots in favourable conditions (Pearson, 2002a, 85). Marsden (1994, 80) estimates the maximum load capacity would have been up to 50 tonnes. Built of oak, it had no keel but featured two broad keel-planks, a stempost with corresponding sternpost and hazel twig caulking for the carvel planking. The mast was supported by a rectangular socket mast-step in the base of which a bronze coin of Domitian was found. Based on dendro-analysis, the vessel was built in the South East of Britain around AD 140. Particularly important to our consideration of the riverine use of such Romano-Celtic vessels on the Medway is the mast, and specifically the type of sail used. Considerable debate has taken place as to whether the Blackfriars 1 vessel's sail was square or was a spirit-sail, especially as the mast was located forward. Marsden (1994, 73) considered both before erring towards a square sail, he saying: *"...since the mast is likely to have been in the region of 12.7m high from the step...the length of the yard would have been about 10m. This would give a sail about 8 x 8m2."*

He points out that this solution would have had limitations when sailing with a beam wind, but notes that the positioning of rows of limber holes in the bottom of the ship indicates that it was designed to sail heeled over at 12 degrees. He concludes that this indicates the Blackfriars 1 ship was expected to sail with a side-wind when necessary.

Milne (2000, 131) has argued that the close similarity in design of the New Guy's House boat, the Blackfriars 1 vessel and others found at Bruges in Belgium, St Peter Port in Guernsey (estimated to have had a crew of three, Rule and Monaghan, 1993) and Barland's Farm in Gwent (this boat was half the size of the Blackfriars boat, though again the same design) indicates a close association with the *Classis Britannica* given the regional footprint of the finds (see discussion above in Chapter 5). However, many other Type B-style vessels of differing styles to these ships have also been found across Europe, for example once again at Laurum, Cologne and Pommeroeul. Lyne (1996) confirms this ubiquity in his analysis of the Richborough ship fittings, with the second most common types originating from carvel-built (planks fitted end-to-end) Romano-Celtic vessels.

Finally and interestingly, Lyne also identifies a third north west European ship building tradition at Richborough, namely Germanic style clinker-built (overlapping-plank) fast rowing vessels. These latter were limited in number and of a late occupation date and he argues they were the vessels of north German foederates, raiders or settlers.

www.ingramcontent.com/pod-product-compliance
Lightning Source LLC
LaVergne TN
LVHW071216210926
848482LV00015B/826